Cell Volume Regulation

The Molecular Mechanism and Volume Sensing Machinery

Cell Volume Regulation

The Molecular Mechanism and Volume Sensing Machinery

Proceedings of the 23rd Taniguchi Foundation Biophysics Symposium held in Okazaki, Japan, 17-21 November 1997

Editor:

Yasunobu Okada

Department of Cellular and Molecular Physiology
National Institute for Physiological Sciences
Okazaki, Japan

1998

ELSEVIER

Amsterdam – Lausanne – New York – Oxford – Shannon – Singapore – Tokyo

International Congress Series No. 1160
ISBN 0-444-82951-2

Published by:
Elsevier Science B.V.
P.O. Box 211
1000 AE Amsterdam
The Netherlands

Library of Congress Cataloging in Publication Data
A catalog record from the Library of Congress has been applied for.

∞ The paper used in this publication meets the requirements of ANSI/NISO Z39.48-1992 (Permanence of Paper).

Printed in The Netherlands.

Dedication

This book is dedicated to the late Toyosaburo Taniguchi who is the former president of the Taniguchi Foundation.

The Taniguchi Foundation Biophysics Symposium
"CELL VOLUME REGULATION: THE MOLECULAR MECHANISM AND VOLUME SENSING MACHINERY"
(November 17-21, 1997)

Preface

A fundamental property of almost all animal cells is their ability to regulate their volume under osmotic perturbation and during cell shape changes. Osmotic perturbation is associated not only with a variety of pathological conditions but also with a large number of cellular physiological activities. Furthermore, it has recently become evident that impairments of volume regulation are closely coupled to apoptotic or necrotic cell death.

In the light of the recent upsurge of interest in cell volume regulation, this publication is extremely timely. This book discusses all of the major areas that are important to an understanding of cell volume regulation. The fundamental question is: How do cells sense and regulate their volume? In answering this question, particular emphasis is placed on our current understanding of the molecular machinery underlying sensing of volume and its regulation.

This book is based on the proceedings of the 23rd Taniguchi Foundation Biophysics Symposium which was organized by me and held in Okazaki, Japan from November 17 to 21 in 1997. Grateful acknowledgment is made to the Taniguchi Foundation for their generous financial support to this symposium. Thanks are also due to staff of the Department of Cellular and Molecular Physiology, National Institute for Physiological Sciences, Okazaki National Research Institutes, for their invaluable assistance.

This publication is supported by the Taniguchi Foundation and by CREST (Core Research for Evolutional Science and Technology) of JST (Japan Science and Technology Corporation).

Yasunobu Okada
Professor
National Institute for Physiological Sciences
Okazaki, Japan

Contributors

N. ANZAI, *Department of Physiology, Kitasato University School of Medicine, Kitasato, Sagamihara 228-8555, Japan*

C. BAZZINI, *Department of Physiology, University of Innsbruck, Fritz-Pregl-Straße 3, A-6020 Innsbruck, Austria*

G.L. BUSCH, *Department of Physiology I, Eberhard-Karls-University of Tübingen, Gmelinstrasse 5, D-72076 Tübingen, Germany*

V. CARDIN *Department Biophysics, Institute of Cell Physiology, National University of Mexico, Ciudad Universitaria, CP 04510, Mexico D.F., Mexico*

L. CHEN, *Physiology Unit, School of Molecular and Medical Biosciences, University of Wales, Cardiff CF1 3US, UK*

T. CHIBA, *Department of Gastroenterology, Kyoto University, Graduate School of Medicine, Yoshida-Konoemachi, Sakyo-ku, Kyoto 606-8507, Japan*

P. DEETJEN, *Department of Physiology, University of Innsbruck, Fritz-Pregl-Straße 3, A-6020 Innsbruck, Austria*

G. DROOGMANS, *Department of Physiology, Campus Gasthuisberg, KU Leuven, B-3000 Leuven, Belgium*

J. EGGERMONT, *Department of Physiology, Campus Gasthuisberg, KU Leuven, B-3000 Leuven, Belgium*

K. ENOMOTO, *Department of Physiology, Shimane Medical University, Enya-machi, Izumo, 693-8501, Japan*

T. FUJITA, *Fourth Department of Internal Medicine, University of Tokyo School of Medicine, Bunkyo-ku, Tokyo 112-0015, Japan*

J. FÜRST, *Department of Physiology, University of Innsbruck, Fritz-Pregl-Straße 3, A-6020 Innsbruck, Austria*

K. FURUYA, *Department Applied Biology, Kyoto Institute of Technology Research Institute, Matsugasaki, Kyoto 606-8585, Japan*

K. FUSHIMI, *Second Department of Internal Medicine, Tokyo Medical and Dental University School of Medicine, Yushima Bunkyo-ku, Tokyo 113-8519, Japan*

M. GSCHWENTNER, *Department of Physiology, University of Innsbruck, Fritz-Pregl-Straße 3, A-6020 Innsbruck, Austria*

E. GULBINS, *Department of Physiology I, Eberhard-Karls-University of Tübingen, Gmelinstrasse 5, D-72076 Tübingen, Germany*

A. HAZAMA, *Department of Cellular and Molecular Physiology, National Institute for Physiological Sciences, Myodaiji-cho, Okazaki 444-8585, Japan*

S. HIROSE, *Department of Biological Sciences, Tokyo Institute of Technology, Nagatsuta, Midori-ku, Yokohama 226-8501, Japan*

T. IKEDA, *Department of Molecular Physiology, National Cardiovascular Center Research Institute, Fujishiro-dai, Suita 565-8565, Japan*

T. IWAMOTO, *Department of Molecular Physiology, National Cardiovascular Center Research Institute, Fujishiro-dai, Suita 565-8565, Japan*

I. IZUMIDA, *Department of Physiology, Kitasato University School of Medicine, Kitasato, Sagamihara 228-8555, Japan*

T.J.C. JACOB, *Physiology Unit, School of Molecular and Medical Biosciences, University of Wales, Cardiff CF1 3US, UK*

H.R. de JONGE, *Department of Biochemistry, Faculty of Medicine and Health Sciences, Erasmus University, 3000 DR Rotterdam, The Netherlands*

K. KAWAHARA, *Department of Physiology, Kitasato University School of Medicine, Kitasato, Sagamihara 228-8555, Japan*

H. KIDA, *Department of Gastroenterological Endoscopy, Kyoto University, Graduate School of Medicine, Yoshida-Konoemachi, Sakyo-ku, Kyoto 606-8507, Japan*

H.K. KIMELBERG, *Departments of Pharmacology/Neuroscience, Division of Neurosurgery, Albany Medical College, Albany, NY 12208, USA*

Y. KOBAYASHI, *Department of Molecular Physiology, National Cardiovascular Center Research Institute, Fujishiro-dai, Suita 565-8565, Japan*

I. KOMURO, *Department of Medicine III, Faculty of Medicine, University of Tokyo, Hongo Bunkyo-ku, Tokyo 113-8655, Japan*

M. KUBO, *Department of Cellular and Molecular Physiology, National Institute for Physiological Sciences, Myodaiji-cho, Okazaki 444-8585, Japan*

H.M. KWON, *Division of Nephrology Department of Medicine, The Johns Hopkins University School of Medicine, 720 Rutland Avenue, Baltimore, Maryland 21205, USA*

F. LANG, *Department of Physiology I, Eberhard-Karls-University of Tübingen, Gmelinstrasse 5, D-72076 Tübingen, Germany*

A. LEPPLE-WIENHUES, *Department of Physiology I, Eberhard-Karls-University of Tübingen, Gmelinstrasse 5, D-72076 Tübingen, Germany*

F. MARUMO, *Second Department of Internal Medicine, Tokyo Medical and Dental University School of Medicine, Yushima Bunkyo-ku, Tokyo 113-8519, Japan*

M. MASUDA, *Department of Structural Analysis, National Cardiovascular Center, Fujishiro-dai, Suita 565-8565, Japan*

T. MITSUIYE, *Department of Physiology, Kyoto University, Graduate School of Medicine, Yoshida-Konoemachi, Sakyo-ku, Kyoto 606-8501, Japan*

T. MIYOSHI, *Department of Cellular and Molecular Physiology, National Institute for Physiological Sciences, Myodaiji-cho, Okazaki 444-8585, Japan*

A. MIWA, *Department of Cellular and Molecular Physiology, National Institute for Physiological Sciences, Myodaiji-cho, Okazaki 444-8585, Japan*

S. MORALES-MULIA, *Department Biophysics, Institute of Cell Physiology, National University of Mexico, Ciudad Universitaria, CP 04510, Mexico D.F., Mexico*

S. MORISHIMA, *Department of Cellular and Molecular Physiology, National Institute for Physiological Sciences, Myodaiji-cho, Okazaki 444-8585, Japan*

U.O. NAGL, *Department of Physiology, University of Innsbruck, Fritz-Pregl-Straße 3, A-6020 Innsbruck, Austria*

N. NAKAMURA, *Department of Biological Sciences, Tokyo Institute of Technology, Nagatsuta, Midori-ku, Yokohama 226-8501, Japan*

T. NAKANISHI, *Department of Kidney and Dialysis, Hyogo College of Medicine, Mukogawa-cho, Nishinomiya, 663-8501, Japan*

H. NAKANO, *Department of Cell Physiology, National Institute for Physiological Sciences, Myodaiji-cho, Okazaki 444-8585, Japan*

B. NILIUS, *Department of Physiology, Campus Gasthuisberg, KU Leuven, B-3000 Leuven, Belgium*

A. NOMA, *Department of Physiology, Kyoto University, Graduate School of Medicine, Yoshida-Konoemachi, Sakyo-ku, Kyoto 606-8501, Japan*

M. OIKE, *Department of Pharmacology, Faculty of Medicine, Kyushu University, Maidashi, Higashi-ku, Fukuoka 812-8582, Japan*

S. OIKI, *Department of Cellular and Molecular Physiology, National Institute for Physiological Sciences, Myodaiji-cho, Okazaki 444-8585, Japan*

Y. OKADA, *Department of Cellular and Molecular Physiology, National Institute for Physiological Sciences, Myodaiji-cho, Okazaki 444-8585, Japan*

H. PASANTES-MORALES, *Department Biophysics, Institute of Cell Physiology, National University of Mexico, Ciudad Universitaria, CP 04510, Mexico D.F., Mexico*

M. PAULMICHL, *Department of Physiology, University of Innsbruck, Fritz-Pregl-Straße 3, A-6020 Innsbruck, Austria*

O. QUESADA, *Department Biophysics, Institute of Cell Physiology, National University of Mexico, Ciudad Universitaria, CP 04510, Mexico D.F., Mexico*

S.A. REES, *University Laboratory of Physiology, Oxford, OX1 3PT, UK*

M. RITTER, *Department of Physiology I, Eberhard-Karls-University of Tübingen, Gmelinstrasse 5, D-72076 Tübingen, Germany*

E.M. RUTLEDGE, *Departments of Pharmacology/Neuroscience, Albany Medical College, Albany, NY 12208, USA*

J.R. SACHS, *Division of Hematology, HSC T15/040, State University of New York, Stony Brook New York 11794-8151, USA*

H. SAKUTA, *Department of Biological Sciences, Tokyo Institute of Technology, Nagatsuta, Midori-ku, Yokohama 226-8501, Japan*

S. SASAKI, *Second Department of Internal Medicine, Tokyo Medical and Dental University School of Medicine, Yushima Bunkyo-ku, Tokyo 113-8519, Japan*

E. SCANDELLA, *Department of Physiology, University of Innsbruck, Fritz-Pregl-Straße 3, A-6020 Innsbruck, Austria*

A. SCHMARDA, *Department of Physiology, University of Innsbruck, Fritz-Pregl-Straße 3, A-6020 Innsbruck, Austria*

M. SHIGEKAWA, *Department of Molecular Physiology, National Cardiovascular Center Research Institute, Fujishiro-dai, Suita 565-8565, Japan*

T. SHIMIZU, *Department of Cellular and Molecular Physiology, National Institute for Physiological Sciences, Myodaiji-cho, Okazaki 444-8585, Japan*

M. SOKABE, *Department of Physiology, Nagoya University School of Medicine, Tsurumai Showa-ku, Nagoya 466-8550, Japan*

Y. SUZUKI, *Department of Biological Sciences, Tokyo Institute of Technology, Nagatsuta, Midori-ku, Yokohama 226-8501, Japan*

I. SZABO, *Department of Physiology I, Eberhard-Karls-University of Tübingen, Gmelinstrasse 5, D-72076 Tübingen, Germany*

K. TAKANO, *Fourth Department of Internal Medicine, University of Tokyo School of Medicine, Bunkyo-ku, Tokyo 112-8655, Japan*

Y. TAKAMITSU, *Department of Kidney and Dialysis, Hyogo College of Medicine, Mukogawa-cho, Nishinomiya, 663-8501, Japan*

K. TAKEMOTO, *Department of Physiology, Nagoya University School of Medicine, Tsurumai Showa-ku, Nagoya 466-8550, Japan*

B.C. TILLY, *Department of Biochemistry, Faculty of Medicine and Health Sciences, Erasmus University, 3000 DR Rotterdam, The Netherlands*

M. TOMINAGA, *Department of Cellular and Molecular Physiology, National Institute for Physiological Sciences, Myodaiji-cho, Okazaki 444-8585, Japan*

T. TOMINAGA, *Department of Cellular and Molecular Physiology, National Institute for Physiological Sciences, Myodaiji-cho, Okazaki 444-8585, Japan*

T. TSUMURA, *Department of Internal Medicine, Kyoto University, Graduate School of Medicine, Shogoin-Kawaramachi, Sakyo-ku, Kyoto 606-8507, Japan*

K. UEDA, *Department of Agricultural Chemistry, Kyoto University, Graduate School of Agricultural, Kitashirakawa-Oiwakemachi, Sakyo-ku, Kyoto 606-8502, Japan*

S. UEDA, *Department of Internal Medicine, Kyoto University, Graduate School of Medicine, Shogoin-Kawaramachi, Sakyo-ku, Kyoto 606-8507, Japan*

T. VOETS, *Department of Physiology, Campus Gasthuisberg, KU Leuven, B-3000 Leuven, Belgium*

S. WAKABAYASHI, *Department of Molecular Physiology, National Cardiovascular Center Research Institute, Fujishiro-dai, Suita 565-8565, Japan*

S. WALDEGGER, *Department of Physiology I, Eberhard-Karls-University of Tübingen, Gmelinstrasse 5, D-72076 Tübingen, Germany*

L. WANG, *Physiology Unit, School of Molecular and Medical Biosciences, University of Wales, Cardiff CF1 3US, UK*

Z. WANG, *Department of Physiology, Kyoto University, Graduate School of Medicine, Yoshida-Konoemachi, Sakyo-ku, Kyoto 606-8501, Japan*

T. van der WIJK, *Department of Biochemistry, Faculty of Medicine and Health Sciences, Erasmus University, 3000 DR Rotterdam, The Netherlands*

S. YAMAMOTO, *First Department of Medicine, Osaka University Medical School, Yamadaoka, Suita 565-0871, Japan*

A. YAMAUCHI, *First Department of Medicine, Osaka University Medical School, Yamadaoka, Suita 565-0871, Japan*

W. YU, *Department of Physiology, Nagoya University School of Medicine, Tsurumai Showa-ku, Nagoya 466-8550, Japan*

Contents

Preface vii

Contributors ix

Part I. Volume Sensor and Volume Regulation

How do red blood cells know how big they are? . . . .
J.R. Sachs 3

Volume sensor anion channel . . . .
Y. Okada, M. Tominaga, T. Tominaga, A. Hazama, T. Tsumura, S. Morishima, S. Oiki, K. Ueda, T. Miyoshi and A. Miwa 15

Platelet endothelial cell adhesion molecule-1 (PECAM-1, CD31) is involved in mechano-signal transduction in endothelial cells . . . .
M. Masuda, M. Osawa, H. Shigematsu, N. Harada, R.B. Lopes, K. Kusano and K. Fujiwara 23

Neural volume regulation of single guinea-pig cardiac myocytes . . . .
T. Mitsuiye and A. Noma 35

Regulatory volume decrease in cultured brain cells: rate limiting factors . . . .
H. Pasantes-Morales, C.V. Morales-Mulia and O. Quesada 43

Cell volume in the regulation of metabolism, cell proliferation and apoptotic cell death . . . .
F. Lang, I. Szabo, A. Lepple-Wienhues, M. Ritter, S. Waldegger and E. Gulbins 49

Part II. Cellular Signalling Associated with Volume Regulation

Activation of cellular signalling pathways by hypotonicity . . . .
B.C. Tilly, T. van der Wijk and H.R. de Jonge 59

Dual calcium response to cell volume change in aortic endothelium . . . .
M. Oike, C. Kimura, T. Koyama and Y. Ito 67

Ionic mechanisms of RVD in A6 cells: SA channel, $[Ca^{2+}]_i$, whole cell currents, and RVD
M. Sokabe, W. Yu and K. Takemoto 71

Cause and roles of mechanical stress in mammary gland . . . .
K. Furuya, H. Nakano and K. Enomoto 85

ATP release from swollen or CFTR-expressing epithelial cells . . . .
A. Hazama, A. Miwa, T. Miyoshi, T. Shimizu and Y. Okada 93

Molecular mechanism of stress-induced cardiac hypertrophy . . . .
I. Komuro *99*

Part III. Ion Channel and Transporter Involved in Volume Regulation

Cell volume regulation: the role of chloride channels . . . .
T.J.C. Jacob, L. Wang and L. Chen *109*

Electrophysiological properties of volume-regulated Cl^- channels in intestinal epithelial cells . . . .
S. Oiki, M. Kubo and Y. Okada *125*

Is ICln a swelling dependent chloride channel? . . . .
M. Gschwentner, J. Fürst, A. Schmarda, C. Bazzini, E. Wöll, M. Ritter, A. Dienstl, E. Scandella, B. Oehl, M. Hohlrieder and M. Paulmichl *131*

Volume-regulated anion channels in vascular endothelium . . . .
B. Nilius, J. Eggermont, T. Voets and G. Droogmans *141*

Molecular machineries of the neonatal rat kidney against osmotic stress . . . .
K. Kawahara, N. Anzai and I. Isumida *151*

Cloning and properties of a hypertonicity-inducible inward rectifier K^+ channel from euryhaline eels . . . .
Y. Suzuki, N. Nakamura, H. Sakuta, K. Takano, T. Fujita and S. Hirose *159*

The Na^+/H^+ exchanger and volume regulation . . . .
S. Wakabayashi, T. Ikeda, Y. Kobayashi, T. Iwamoto and M. Shigekawa *167*

Part IV. Organic Osmolyte Transporter and Water Channel Involved in Volume Regulation

Transcriptional regulation of transporters for compatible osmolytes by hypertonicity
H.M. Kwon *175*

Urinary concentrating defect and organic osmolytes . . . .
T. Nakanishi, A. Yamauchi, S. Yamamoto and Y. Takamitsu *181*

Hypotonic and high K^+ media swelling-induced release of excitatory amino-acids from brain astrocytes. . . . .
H.K. Kimelberg and E.M. Rutledge *189*

Structure and regulation of the aquaporin water channels . . . .
K. Fushimi, S. Sasaki and F. Marumo *199*

Water movement during cell volume regulation. . . . .
S. Morishima, H. Kida, S. Ueda, T. Chiba and Y. Okada *209*

Index of authors *213*

Part I
Volume Sensor and Volume Regulation

Cell Volume Regulation: The Molecular Mechanism and Volume Sensing Machinery
Y. Okada, editor.

How do red blood cells know how big they are?

John R. Sachs

Department of Medicine, State University of New York at Stony Brook, Stony Brook, N.Y. 11794-8151

INTRODUCTION

Red cells, like other cells, are able to maintain a constant volume throughout their life span. For mammalian red cells, many of which are biconcave discs, a constant relation between volume and surface area is critical; if the cells were to swell and become spherical, they would lose their deformability and be removed from the circulation. Although most mammalian red cells are relatively impermeable to cations, nevertheless they have cation transporters similar to those of more complicated cells, and the activity of these transporters varies considerably from individual to individual. Over the course of 120 days, operation of these transporters would be expected to result in significant change in cation content and therefore volume. Cation transport processes in red cells are much too slow to permit study of volume regulation in steady state isosmotic conditions, and all of what is known about this subject comes from studies in which cell volume is altered more or less drastically.

Na-H exchangers and Na, K, 2Cl cotransporters activated by cell shrinkage have been described in red cells from a variety of species, and K-Cl cotransporters activated by cell swelling are present in most red cells; in fact, the existence of these cotransport mechanisms and their response to cell volume were first described by Kregenow in duck red blood cells many years ago. The characteristics of the transporters have been extensively described and will not be recounted here.

Since they are simple structures and easily manipulated, red cells are attractive experimental objects for asking questions about how volume is sensed and how the sensor communicates with the transporters. There are some possible mechanisms of volume sensing which can be immediately discarded. Human red cells and the red cells of many other species are biconcave discs. It would be necessary for them to swell about 1.8 times before they become spheres and therefore capable of developing membrane tension. Since changes in cell volume much less than this activate K-Cl cotransport, mechanisms in which membrane stretch plays a role can be discarded. Red cells are nearly devoid of intracellular Ca, and nearly all studies of volume regulation have been done in Ca-free solutions. Models of volume sensing which involve any increase in intracellular Ca can be discarded. Mammalian red cells do not contain DNA and therefore mechanisms of volume regulation which involve gene activation are not possible. Many of the common signal transduction pathways present in more complicated cells do not operate in red blood cells, and therefore their role in volume regulation in these cells can be disregarded *a priori.*

SENSING OF VOLUME BY CHANGE IN INTRACELLULAR SOLUTE CONCENTRATION

Currently, there is an elegant model of cell volume sensing which has been widely accepted. Jennings and Al-Rohil (1) showed that, when rabbit red cells were suddenly swollen, K-Cl cotransport increased, but the increase occurred with a considerable time lag. On the other hand, when swollen cells were suddenly shrunk, decrease in K-Cl cotransport rate occurred without a lag. To interpret these findings, the authors applied a principle from relaxation kinetics. For the reaction:

$$T_A \underset{k_{-1}}{\overset{k_1}{\rightleftharpoons}} T_I$$

The rate of transformation from T_I (inactive cotransporter) to T_A (active cotransporter) after a sudden perturbation is ($k_1 + k_{-1}$). From the existence of a lag in the activation of cotransport after a sudden increase in cell size, the authors concluded that swelling activates by reducing k_1, rather than by increasing k_{-1}. It was soon reported that okadaic acid, an inhibitor of serine/threonine phosphatases, inhibited K-Cl cotransport and its activation by swelling (2,3). Okadaic acid inhibited cotransport activation by preventing the dephosphorylation of T_I, the inactive phosphorylated cotransporter. The rate constant k_1, which is reduced by cell swelling, must therefore describe the activity of a serine/threonine kinase. The kinase is volume sensitive, and its activity is decreased by cell swelling.

Soon after these results were obtained, Minton, Colclasure and Parker (4) proposed that cell volume is sensed by macromolecular crowding. Concentration or dilution of macromolecules in solutions results in nonlinear thermodynamic behavior. In solution, macromolecules occupy a finite volume which is not available for solution of other molecules. The volume from which a small molecule is excluded is not much larger than the volume physically occupied by the macromolecules since small molecules are able to fit into all the intermolecular spaces. Macromolecules are excluded from a much larger volume since they are too large to fit into much of the space between macromolecules. Since macromolecules have access to only a fraction of the solvent volume, their concentration, and therefore their thermodynamic activity, is greater than expected if concentration is calculated from total solvent volume. Moreover, further concentration of the macromolecular solution increases the excluded volume more rapidly than the volume physically occupied by the macromolecules increases, and the thermodynamic activity of the macromolecules is an exponential function of its apparent concentration. Evidence has been presented consistent with the proposal that macromolecular crowding is the sensor of cell volume in dog red blood cells.

The two findings complement each other. It is supposed that the activity of a serine/threonine kinase is controlled by the concentration of macromolecules. When cells are shrunken, macromolecules are crowded, protein kinase activity is increased, the K-Cl cotransporter is phosphorylated and inactivated, and the Na-H exchanger phosphorylated and activated. When cells are swollen, the reverse sequence occurs. The theory is elegant and provides an intellectually satisfying explanation for many of the findings about volume regulation in red cells of many species (5). However, it is not the whole story.

K-Cl COTRANSPORT IN HUMAN RED CELL GHOSTS

We have examined swelling activated K-Cl cotransport in human red cell ghosts. In order to understand the results, it is necessary to know how the ghosts are prepared. Red cells are run into a column containing BioGel beads with a MW exclusion of about one million D which are equilibrated with a hypotonic solution (about 20 mOsm/kg H_20). The cells hemolyze on the column and all small molecules and soluble proteins are absorbed to the beads. The eluted ghosts are free of ATP, Na, K, hemoglobin, etc. The ghosts are then suspended in a solution whose composition is that which we wish the ghosts to have during the measurement of K-Cl cotransport. The ghosts are incubated at 37°C for 30 to 60 minutes in order to reseal them, washed, and used for the measurement of K influx in a solution of known composition. We have demonstrated that the characteristics of K-Cl cotransport in these ghosts are exactly the same as its characteristics in intact red cells of humans and other species (6).

By altering the concentration of solutes in the solutions in which the ghosts are resealed, we are able to prepare ghosts which vary in volume, but have identical concentrations of intracellular solutes when they are all suspended in solutions of the same osmolality. In these ghosts, K-Cl cotransport increases with ghost volume even though solute concentration is constant and the ghosts are virtually free of protein (Fig. 1). Macromolecular crowding cannot be the volume sensor in these ghosts since there are no macromolecules to be crowded or to do the crowding. When ghosts are swollen, K-Cl cotransport rate increases immediately with

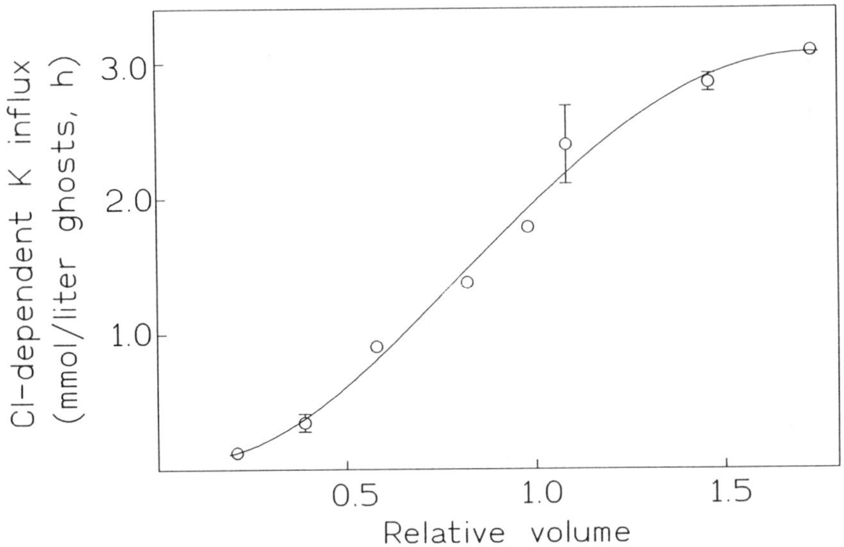

Figure 1. Ouabain-insensitive, chloride-dependent K influx vs. the volume of the ghosts relative to their volume at resealing.

no detectable time lag, and when swollen ghosts are reshrunk, K-Cl cotransport rate decreases immediately with no detectable lag. This is different from the situation with intact cells, and eliminates one of the supports for the hypothesis that inhibition of a kinase activates K-Cl cotransport in response to cell swelling.

In other experiments, we exposed ghosts to okadaic acid at the beginning of the resealing incubation, midway through the resealing incubation, and during the measurement of cotransport rate alone. After resealing, the ghosts were swollen and cotransport measured (Table 1). When added at the beginning of the resealing incubation, okadaic acid nearly completely inhibited K-Cl cotransport in swollen ghosts; when added halfway through the resealing incubation, cotransport was partially inhibited; and when present only during the cotransport measurement, little inhibition was seen. In these ghosts, dephosphorylation of the relevant serine/threonine group occurs during resealing and before ghost swelling occurs. Nevertheless, even though the relevant group is dephosphorylated, swelling activates K-Cl cotransport. We believe the relevant protein kinase is cytosolic and lost during preparation of the ghosts, but the phosphatase is at least in part membrane bound.

Table 1
Inhibition of ouabain-insensitive, chloride-dependent K influx at varying times after exposure to okadaic acid.

Okadaic acid during resealing	Okadaic acid during influx measurement	K influx (mmol/liter ghosts, h)	Fractional inhibition
None	None	4.88	—
	0.25 μM	4.27	0.12
0.25 μM 30 minutes after start of resealing	None	3.20	0.34
	0.25 μM	3.04	0.38
0.25 μM at start of resealing	None	0.57	0.88
	0.25 μM	0.59	0.88

Finally, we found that ghosts prepared free of ATP increase their K-Cl cotransport rate when swollen (although the magnitude of the increase is less than when ghosts containing ATP are used), and when the swollen ghosts are re shrunk, cotransport rate decreases (Table 2). Since there is no ATP in these ghosts, ATP is not necessary for swelling activated increase in cotransport rate. Moreover, since the swollen ATP-free ghosts decrease their cotransport rate when reshrunk, phosphorylation cannot be responsible for the decrease in the cotransport rate when swollen ATP-free ghosts are reshrunk (7).

Table 2
Ouabain-insensitive influx in ATP-free ghosts.

Ghosts	K influx (mmol/liter ghosts, h)
Swollen ⟶ swollen	2.55
Swollen ⟶ shrunken	0.81
Shrunken ⟶ swollen	2.69
Shrunken ⟶ shrunken	0.93

The situation is complicated in that, in intact red blood cells, ATP increases cotransport rate at all cell volumes. In ghosts, we have found that the substrate for this stimulation is MgATP, that stimulation is inhibited by the tyrosine kinase inhibitor genistein and that, in ATP-free ghosts, K-Cl cotransport is activated by vanadate, a tyrosine phosphatase inhibitor. Stimulation of K-Cl cotransport by ATP is probably mediated by a tyrosine kinase (7). Since volume stimulation is apparent even in ATP-free ghosts, tyrosine phosphorylation is not necessary for such stimulation.

Consideration of the results described above led us to the conclusion that phosphorylation and dephosphorylatioin events are not in the direct signal transduction pathway by which the sensor of cell volume communicates with the cotransporter. Results entirely compatible with ours have been reported by Dunham and his collaborators (8) using intact LK sheep red blood cells and inside-out-vesicles obtained from them.

We have examined a number of other second messengers (cyclic AMP, cyclic GMP, G proteins, NO, eicosanoids) to determine whether, however unlikely, they are involved in the signal transaction pathway (6,7). We could find no evidence for such involvement. We came to the conclusion that, in ghosts, the volume sensor communicates directly with the cotransporter.

Although phosphorylation is not in the signal transduction pathway , there is no doubt that inhibition of a serine/threonine phosphatase reduces volume-stimulated K-Cl cotransport. Either there are two pathways for volume sensing and cotransport stimulation, or the observation must be accounted for in some other way. We have considered the possibility that phosphorylation is conformation dependent, i.e., that swelling changes the conformation of the cotransporter from an inactive (T_I) to an active (T_A) form, and that serine/threonine phosphorylation of the inactive, but not the active, conformation of the cotransporter can occur:

$$T_A \underset{k_{-1}}{\overset{k_1}{\rightleftharpoons}} T_I \underset{k_{-2}}{\overset{k_2\,ATP}{\rightleftarrows}} T_IP$$

where $k_1 = k_1\, f(V/V_o)$, and $k_{-1} = k_{-1} / f(V/V_o)$. We have separated the phosphorylation step into two separate steps since phosphorylation is not a reversible reaction, but is, in fact carried out by two separate enzymes. The step is likely to be even more complicated since it involves binding of the kinase, phosphorylation, release of the kinase, binding of the phosphatase, dephosphorylation, and release of the phosphatase. Whether or not such a mechanism would predict a lag in the activation of cotransport by swelling and no lag in the deactivation of cotransport on shrinking is not intuitively obvious; the kinetics are likely to be very complex. It should be pointed out that conformation dependent phosphorylation has been described with other enzymes (9,10).

Although swelling activates K-Cl cotransport in the absence of macromolecules, it will be necessary to reconcile the findings with ghosts with the clear demonstration of an effect of macromolecular crowding in dog red cells. In red blood cells, the principal macromolecule is hemoglobin, and it is known that K-Cl cotransport rate is modified by mutant hemoglobins such as sickle hemoglobin (11). It may be possible for hemoglobin to interact directly with the volume sensor to modulate its effects. By increasing the thermodynamic activity of hemoglobin, macromolecular crowding might enhance this effect. In this formulation, macromolecular crowding is not the primary volume sensor, but a modulator of some other sensor, and it provides another fine control of the volume regulatory mechanism.

ROLE OF THE CYTOSKELETON IN VOLUME SENSING

In red cell ghosts, the entire mechanism of volume-sensing and cotransporter activation resides in the cell membrane. We have performed some experiments to evaluate the role of cytoskeletal proteins in this process.

In many preparations, actin has been implicated in cell volume regulation. In red cells, actin exists as an actin monofilament 10-14 actin monomers long. The monofilament is capped at both the barbed and pointed ends, and it binds up to six spectrin molecules along its length. Actin filaments in red cells and their ghosts are, therefore, very stable and it seemed unlikely that actin depolymerization is involved in cell volume sensing or signal transduction in red cell ghosts. Nevertheless, we performed experiments in which ghosts were swollen in the presence of high concentrations of cytochalasin (which prevents addition of actin monomers to F actin filaments) or latrunculin (which binds G actin with high affinity). Cotransport rate in the swollen ghosts was measured. The ghosts were then reshrunk and cotransport rate again measured; shrinking of the ghosts reduced the cotransport rate to the same extent whether or not cytochalasin or latrunculin was present. If the depolymerization of actin filaments was responsible for the elevated cotransport rate of swollen ghosts, cytochalasin should have prevented reformation of the filaments when the ghosts were reshrunk, and therefore prevented the reduction of the cotransport rate. On the other hand, if G actin molecules released from actin filaments signal cell swelling, latrunculin, by binding G actin, should either have prevented swelling activation of cotransport or prevented the reduction of cotransport rate when the swollen ghosts were reshrunk. Since neither of these predicted behaviors was observed we concluded that actin depolymerization is not involved in regulation of K-Cl cotransport by ghost swelling.

The cytoskeleton is attached to the red cell membrane by two proteins: ankyrin , which binds β spectrin to band 3 protein, and band 4.1 protein which binds spectrin to glycophorin C and band 3 protein. Before resealing, we incubated ghosts with trypsin for varying periods of time and then added a trypsin inhibitor. Trypsin significantly digested both ankyrin and band 4.1 protein, but K-Cl cotransport rate in shrunken or swollen ghosts was not altered, nor was the response of the cotransport rate to ghost swelling.

It is known that heating red blood cells or their ghosts to approximately 49 ° C irreversibly denatures ß spectrin without any apparent denaturation of any other cytoskeletal or membrane protein. As a result of the denaturation of ß spectrin, formation of spectrin tetramers from spectrin dimers is defective. Orlov et al. (12) reported that heating rat red cells had no effect on Na-H exchange rate or Na, K, 2Cl cotransport rate in cells of normal volume, but prevented the increase in the transport rates when the cells were shrunk. On the other hand, heating greatly increased K-Cl cotransport rate at normal cell volume, but prevented further increase on cell swelling. The results indicate that denaturation of spectrin shifts the set point of the volume sensor to lower cell volumes. We found that heating red cell ghosts reduced K-Cl cotransport rate slightly in shrunken ghosts, but greatly increased it in swollen ghosts; the results are similar to those obtained with rat red cells. In human ghosts the stimulation of K-Cl cotransport rate is reversible even though the denaturation of ß spectrin is not; the increased cotransport rate in swollen heated ghosts returns to low levels when the swollen ghosts are reshrunk. The possible significance of these results will be discussed below, but it is an indication that some interaction of the cotransporter with the cytoskeleton may alter the cotransport rate in response to changes in cell volume.

EFFECTS OF ORGANIC CATIONS ON COTRANSPORT RATE

In red cell ghosts, as in other red cell preparations which have been investigated, divalent cations such as Mg^{2+} inhibit K-Cl cotransport rate at high concentrations in both swollen and shrunken ghosts (13). We have found that soluble polycations (spermine, neomycin, methylglyoxal) and amphipathic cations (sphingosine, tetracaine, trifluoperazine) which insert into hydrophobic regions, especially the phospholipid bilayer membrane, inhibit K-Cl cotransport in both shrunken and swollen ghosts (14). The divalent cations and soluble polycations (but not the amphipathic cations) could inhibit cotransport by accumulating in the diffuse double layer adjacent to the inner phospholpid bilayer (the inner bilayer of the red cell membrane, like that of other cells, is negatively charged since it contains high concentrations of phosphatidylserine), or, alternatively, by binding to negatively charged molecules in the cell membrane. A method for distinguishing between the two possible mechanisms was suggested by the work of McLaughlin et al. (15). Ethylenediamine ($H_2N\text{-}(CH_2)_2\text{-}NH_2$) both accumulates in the diffuse double layer and binds to charges such as phosphatidylserine. On the other hand, dimethonium ($(CH_3)_3\text{-}N\text{-}(CH_2)_2\text{-}N\text{-}(CH_3)_3$), in which the positively charged nitrogens are shielded by methyl groups, accumulates in the diffuse double layer, but does not bind to phosphatidylserine. Ethylenediamine inhibits K-Cl cotransport in swollen ghosts, although with a low affinity, but dimethonium does not inhibit at all. Similarly, $H_2N\text{-}(CH_2)_{10}\text{-}NH_2$, diaminodecane, inhibits with a relatively high affinity, but $(CH_3)_3\text{-}N\text{-}(CH_2)_{10}\text{-}N\text{-}(CH_3)_3$, decamethonium, inhibits only slightly, and with a very low affinity. It can be concluded that polyvalent cations inhibit by binding to negative charges and neutralizing them.

There is an extensive literature concerned with the quantitative aspects of the interaction of polyvalent ligands with mobile receptors in cell membranes. If the receptors (negative charges)

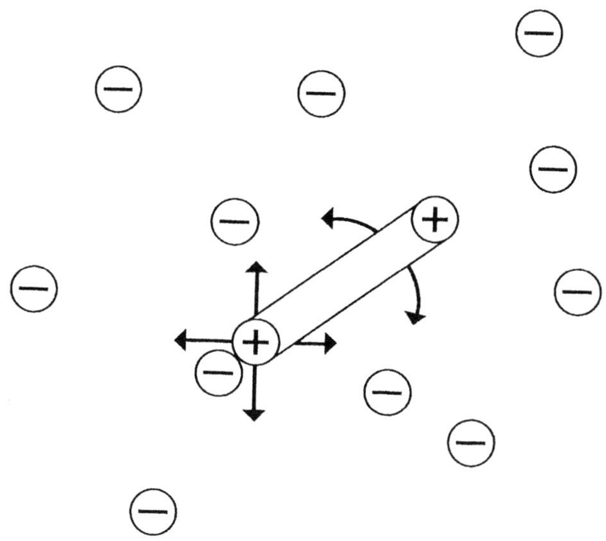

Figure 2. Interaction of a polycation with mobile negative charges in a cell membrane. Binding of the first positive charge to a membrane negative charge reduces the translational mobility of the polycation to two dimensions and reduces its rotational mobility. If the positive charges of the polycation are separated by a distance greater than the distance separating the mobile charges at their closest approach, then binding of the positive charge greatly increases the affinity of the second positive charge for a second mobile negative charge , and the affinity of the whole molecule for negative charges is greater than that of a comparable molecule in which the charges are separated by a distance comparable to or less than that of the negative charges at their distance of closest approach.

are mobile, then the polyvalent ligand (polycation) will behave as a single muliply charged ion and bind with relatively low affinity if the multiple charges on the polycation are separated by a distance small compared to the average separation of the mobile charges. On the other hand, if separation of the charges in the polycation is comparable to or greater than the separation of the mobile membrane charges, then the reduction of dimensionality after binding of the first positive charge of the polycation to a mobile membrane charge greatly increases the affinity of the second positive charge for a second negative membrane charge (Fig. 2). We compared the efficiency of ethylenediamine as an inhibitor of K-Cl cotransport with that of a diamine in which the amino groups were separated by a greater distance (diaminodecane). Diaminodecane was a much more effective inhibitor. We also compared the efficiency of spermine ($H_2N-(CH_2)_3-NH-(CH_2)_4-NH-(CH_2)_3-NH_2$) with that of a molecule which represents half of the spermine molecule ($H_2N-(CH_2)_3-NH-CH_3$). Spermine was a much more efficient inhibitor. We believe these results indicate that the negative charges which the polycations neutralize are located relatively close together and are relatively mobile. These results can be understood if the mobile negative charges are provided by phosphatidylserine of the inner bilayer membrane.

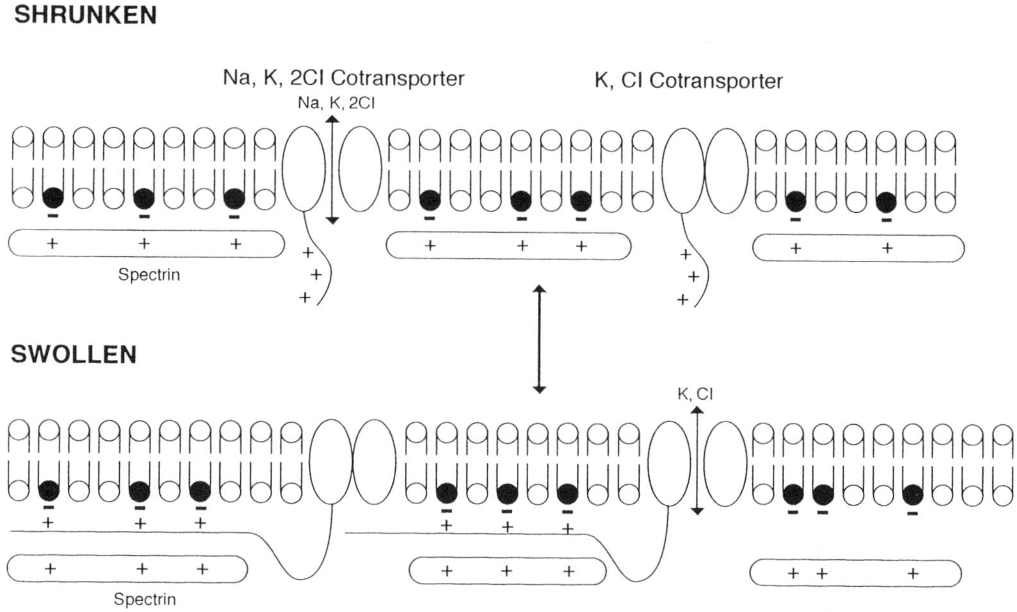

Figure 3. A proposal for the mechanism of volume sensing by red blood cell ghosts. For description, refer to the text.

A HYPOTHESIS FOR THE MECHANISM OF VOLUME SENSING BY HUMAN RED CELL GHOSTS

When we suggest that, in red cell ghosts, no second messenger or phosphorylation-dephosphoylation event is involved in the transmission of the signal of cell volume change from the volume sensor to the cotransporter, we are frequently challenged to propose a mechanism which might be operating. We have therefore developed a hypothesis which incorporates a role for the two manipulations which we have found actually modify the response of the cotransporter to cell swelling, and also accounts for the observation that, in dog and duck red cells, manipulations which modify the activity of the K-Cl cotransporter frequently have a reciprocal and proportional effect on the Na-H exchanger and the Na, K, 2Cl cotransporter.

Fig. 3 illustrates our working hypothesis. We imagine that, in shrunken ghosts, spectrin is closely associated with the inside of the phospholipid bilayer and neutralizes the negative charges of phosphatidylserine (it is known that spectrin associates with phospholipid bilayer membranes containing phosphatidylserine). When ghosts swell, spectrin moves away from the membrane, and cytoplasmic regions of the Na, K, 2 Cl cotransporter and of the K-Cl cotransporter bind to the uncovered negative charges. Both cotransporters undergo a conformational change; Na, K, 2Cl cotransport is inactivated and K-Cl cotransport is activated. The model owes much to the mechanism found to be responsible for the regulation of the ClC-2 (16) channel by cell volume, and it has the virtue of suggesting a way by which membrane expansion might serve as a volume sensor. The model is of course not unique: transporters in which volume sensing is a property of the transporter itself are also consistent with the evidence. The model does, however, guide us in planning new experiments.

CONCLUSION

Red blood cell ghosts offer an attractive preparation for studying the mechanism by which cell volume is sensed. The ghosts are relatively simple structures and they lack many of the candidates for volume sensor which have been identified in other cells. Candidates for volume sensor identified in ghosts can be evaluated in dog and duck red blood cells for reciprocal effects on swelling and shrinkage activated transporters; such reciprocal effects will almost certainly be a characteristic of a volume sensor.

It is, of course, possible that volume sensing mechanisms used in a specialized cell such as a red blood cell are not used by more complicated cells, and it is possible that volume sensors of ghosts are different from those of red cells. It has been pointed out that the response of ghosts to swelling is quantitatively different from the response of intact red cells. Moreover, red cells and their ghosts respond to swelling by activating cotransporters rather than separate K and Cl channels. However, whatever the quantitative differences between the response of ghosts and the response of other cells, including red cells, it must be emphasized that when ghosts are swollen, they increase K influx, the increase is Cl-dependent, and the activation is reversed when swollen ghosts are reshrunk. Whether or not the volume sensor of ghosts is the same as that of other cells, regulation of K-Cl cotransport in ghosts is certainly a process which involves a volume sensor.

The experimental work described in this paper was supported in part by a United States Public Health Service grant DK-19185.

REFERENCES

1. Jennings, M.L., and N. Al-Rohil. (1990). Kinetics of activation and inactivation of swelling-stimulated K^+/Cl^- transport. The volume-sensitive parameter is the rate constant for inactivation. Journal of General Physiology. 95:1021-1040.

2. Jennings, M.L., and R.K. Schulz. (1991). Okadaic acid inhibition of KCl cotransport. Evidence that dephosphorylation is necessary for activation of transport by either cell swelling or N-ethylmaleimide. Journal of General Physiology. 97:799-818.
3. Kaji, D.M., and Y. Tsukitani. (1991). Role of protein phosphatase in activation of KCl cotransport in human erythrocytes. American Journal of Physiology. 260:C178-C182.
4. Minton, A.P., G.C. Colclasure, and J.C. Parker. (1992). Model for the role of macromolecular crowding in regulation of cellular volume. Proceedings of the National Academy of Sciences (USA). 89:10504-10506.
5. Parker, J.C. (1993). In defense of cell volume? American Journal of Physiology. 265:C1191-C1200.
6. Sachs, J.R. (1988). Volume-sensitive K influx in human red cell ghosts, Journal of General Physiology. 92:685-711.
7. Sachs, J.R,. and D.W. Martin. (1993). The role of ATP in swelling-stimulated K-Cl cotransport in human red cell ghosts. Phosphorylation-dephosphorylation events are not in the signal transduction pathway. Journal of General Physiology. 102:551-573.
8. Dunham, P.B., J. Klimczak, and P.J. Logue. (1993). Swelling activation of K-Cl cotransport in LK sheep erythrocytes: A three state process. Journal of General Physiology. 101:733-766.
9. Feschenko, M.S., and K.J. Sweadner. (1994). Conformation-dependent phosphorylation of Na,K-ATPase by protein kinase A and protein kinase C. Journal of Biological Chemistry. 269:30436-30444.
10. Adler, V., M.R. Pincus, T. Minamoto, S.Y. Fuchs, M.J. Bluth, P.W. Brandt-Rauf, F.K. Friedman, R.C. Robinson, J.M. Chen, X.W. Wang, C.C. Harris, and Z. Ronai. (1997). Conformation-dependent phosphorylation of p53. Proceedings of the National Academy of Sciences. (USA) 94:1686-1691.
11. Oliverieri, O., D.Vitoux, F. Galacteros, D. Bachir, Y. Bloquit, Y. Beuzard, and C. Brugnara. (1992). Hemoglobin variants and the activity of the (K + Cl) cotransport system in human erythrocytes. Blood. 79:793-797.
12. Orlov, S.N., I.A. Kolosova, E.J. Cragoe, T.G. Gurlo, A.A. Mongin, S.L. Aksentsev, and S.V. Konev. (1993). Kinetics and peculiarities of thermal inactivation of volume-induced Na^+/H^+ exchange, $Na^+,K^+,2Cl^-$ and K^+, Cl^- cotransport in rat erythrocytes. Biochimica et Biophysica Acta. 1151:186-192.
13. Lauf, P.K., J. Bauer, N.C. Adragna, H. Fujise, A.M.M. Zode-Oppen, K.H. Ryn, and E. Delpire. (1992). Erythrocyte K-Cl cotransport. Properties and regulation. American Journal of Physiology. 263:C917-C932.
14. Sachs, J.R. (1994). Soluble polycations and cationic amphiphiles inhibit volume sensitive K-Cl cotransport in human red cell ghosts. American Journal of Physiology, 266:C997-C1005.
15. McLaughlin, A., W.-K. Eng, G. Vaio, T. Wilson, and S. McLaughlin. (1983). Dimethonium, a divalent cation that exerts only a screening effect on the electrostatic potential adjacent to negatively charged phospholipid bilayer membranes. Journal of Membrane Biology. 76:183-193.
16. Gründer, S., A. Thiemann, M. Pusch, and J. Jentsch. (1992). Regions involved in the opening of the ClC-2 chloride channel by voltage and cell volume. Nature. 360:759-762.

Volume sensor anion channel

Y. Okada[a], M. Tominaga[a], T. Tominaga[a], A. Hazama[a], T. Tsumura[b], S. Morishima[a], S. Oiki[a], K. Ueda[c], T. Miyoshi[a] and A. Miwa[a]

[a]Department of Cellular and Molecular Physiology, National Institute for Physiological Sciences, Okazaki 444-8585, Japan

[b]Department of Internal Medicine, Kyoto University Faculty of Medicine, Kyoto 606-8501, Japan

[c]Department of Agricultural Chemistry, Kyoto University, Kyoto 606-8501, Japan

SUMMARY

In most vertebrate cell types (other than erythrocytes), the regulatory volume decrease (RVD) is mainly attained by activation of K^+ and Cl^- channels after osmotic cell swelling. Volume-sensitive outwardly rectifying anion channels have been shown to be involved in RVD in a large variety of cells. Although a number of candidates, which include P-glycoprotein (PGP), pI_{Cln}, ClC2 and ClC3, for the molecular identity have been listed, none of them has been completely susceptible of proof. In human epithelial Intestine 407 cells which have been proved to express mRNA of PGP, pI_{Cln} and ClC2, our data are not fully compatible with the hypothesis that either protein alone is the channel identity. Recent our studies have shown that activation of swelling-induced Cl^- currents are not associated with membrane stretch but with unfolding of membrane invaginations or folds. Thus, the channel can be called volume expansion-sensing outwardly rectifying or volume sensor (VSOR) Cl^- channel.

INTRODUCTION

A number of transport pathways have been proposed to participate in the processes of RVD and regulatory volume increase (RVI), as shown in Fig. 1. A large variety of cell types other than erythrocytes can attain RVD by activation of K^+ and Cl^- channels [1-3]. Swelling-activated Cl^- channels exhibit common phenotypic characteristics such as volume sensitivity, moderate outward rectification, intracellular ATP requirement, intermediate single-channel conductance, low-field anion selectivity [4].

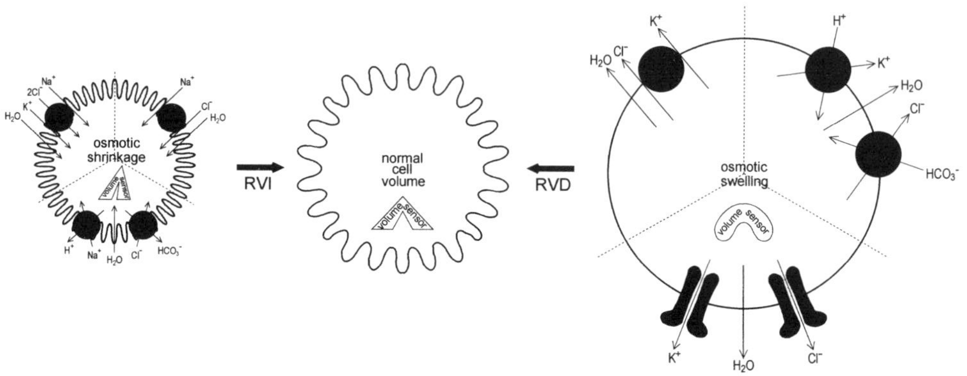

Figure 1. Schematic model of ionic mechanisms for RVD and RVI. Three different systems have been reported for either RVD or RVI mechanism depending on the cell types.

MOLECULAR IDENTITY

Much molecular biological work has been performed to identify the protein of volume-sensitive outwardly rectifying (VSOR) anion channel. However, no perfect proof positive of the molecular identity has been offered to the candidate proteins including P-glycoprotein (PGP), pI_{Cln}, ClC2 and ClC3. Also, the molecular mechanism of swelling-induced activation of VSOR Cl^- channel has not as yet been clarified. Main questions of our current investigations are as follows: What is the molecular identity of VSOR Cl^- channel? How does osmotic swelling activate the VSOR Cl^- channel?

The minimum requirements or *criteria* for identification of the channel protein should consist of three lines of evidence: 1) Cells transfected with the gene of this particular candidate protein do exhibit phenotypic characteristics identical to the VSOR Cl^- current, 2) the cells exhibiting the VSOR Cl^- current do express endogenously the candidate mRNA and protein, and 3) abolition of expression of the candidate protein does completely suppress the endogenous VSOR Cl^- current.

PGP was proposed to switch from an active drug transporter to the VSOR Cl^- channel upon osmotic swelling [5, 6]. However, a number of studies have shown that swelling-activated Cl^- currents were recorded from cells in which PGP expression was undetectably low [7], the fact opposing to the above *criterion*-2. In Intestine 407 cells endogenous expression of PGP was confirmed by Northern blot analysis, RT-PCR, Western blot analysis and immunostaining [8]. However, abolition of the PGP expression by antisense oligonucleotides failed to affect VSOR Cl^- currents [8], the fact being against the *criterion*-3. Recently, Vanoye et al. [9] raised a possibility that endogenous PGP expression in Intestine 407 cells may had been eliminated after detaching the cells from the substrate, based on such observations in polarized hepatocytes [10]. However, our immunostaining study demonstrated significant PGP expression in a dedifferentiated and non-polarized cell line Intestine 407, even 3 h after isolation (A. Hazama & Y. Okada, unpublished observations). Furthermore, treatment with anti-PGP monoclonal antibodies such as C219, MRK16, MRK17 and UIC2 never affected the VSOR Cl^- channel current [8]. Vanoye et al. [9] suggested that the time period of treatment with monoclonal antibodies in our study had been too short (~2 min). However, our studies were done by treating the cells with antibodies against PGP for 10-30 min, though we did not definitely note this time period in our paper [8].

pI_{cln} cloned from MDCK cells has shown to exhibit constitutively activated Cl^- currents, when expressed in *Xenopus* oocytes, by Paulmichl et al. [11]. Although the current has properties largely similar to those of VSOR Cl^- current, several distinct differences between them were also noted by Voets et al. [12]. A most outstanding property of pI_{Cln}-associated current is its sensitivity to extracellular cAMP and cGMP [11]. In Intestine 407 cells, however, VSOR Cl^- currents were not inhibited by extracellular cyclic nucleotides [13], although we found endogenous expression of pI_{Cln} mRNA in the cell by RT-PCR, as shown in Fig. 2. Only partial suppression of VSOR Cl^- currents by treatment with antisense oligonucleotides against pI_{Cln} has been shown in NIH/3T3 cells [14]. Thus, the *criteria*-1 and -3 have not fully been fulfilled for pI_{Cln}.

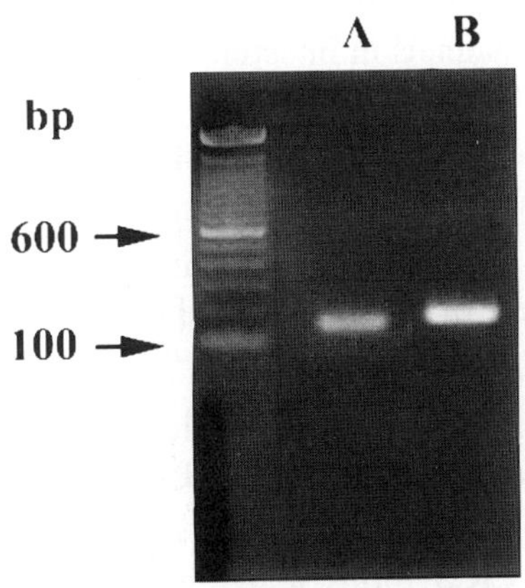

Figure 2. RT-PCR analysis of ClC2 mRNA (*lane* A) and pI_{Cln} mRNA (*lane* B) from Intestine 407 cells. The first *lane* represents size markers. ClC2-specific sequences were amplified by using the primers CCTTCAGTGCCTTCATCTTCC and CCGAAGCCACTAGCAATACC, and pI_{Cln}-specific primers were TGGAGGCAATGTTCACTGC and GGCCTTCTGCTGTTAGATGG.

Since a member of voltage-dependent Cl^- channel family, ClC2, was found to be activated by cell swelling [15], there exists a possibility that ClC2 forms the VSOR Cl^- channel. The ClC2 mRNA is known to be ubiquitously expressed [16]. Our RT-PCR study indicated that Intestine 407 cells also possess the mRNA, as shown in Fig. 2. The properties of VSOR Cl^- currents were, however, totally different from those of ClC2 Cl^- currents recorded in cells which were transfected with the ClC2 gene, as summarized by Okada [4]. ClC2G, which is the rabbit isoform of ClC2, showed Cl^- currents, properties of which are different from ClC2 but rather closer to VSOR, when incorporated in planar lipid bilayers [17]. Its activation by acidic pH or by protein kinase A is, however, distinct from the VSOR Cl^- current. Although the accumulated data do not fit to the *criterion*-1, at present, there remains a possibility that heteromultimeric channel formation between ClC2 and other ClC members may give rise to currents identical to VSOR Cl^- currents, because the ClC2/ClC1 heteromultimer was shown to exhibit Cl^- currents distinct from either ClC1 or ClC2 [18]. In order to draw a definite conclusion, thus, future examination regarding the effects of abolition of endogenous ClC2 expression on swelling-induced Cl^- currents would be essential (for the *criterion*-3).

Recently, a new ClC3 hypothesis has been put forward by Duan et al. [19]. When ClC3 cloned from cardiac myocytes was stably expressed in NIH/3T3 cells, outwardly rectifying Cl^- currents became basally activated, and those activity was further enhanced by osmotic swelling. The current properties were similar to those of VSOR Cl^- currents with respect to inactivation kinetics at positive potentials, anion selectivity of $I^- > Cl^-$, DIDS sensitivity, tamoxifen-sensitivity and extracellular ATP sensitivity. Thus, it is possible that ClC3 alone is the molecular identity of VSOR Cl^- channel in guinea pig cardiac myocytes. However, significant basal activity in non-swollen cells and inhibition by activation of protein kinase C are distinct from VSOR Cl^- channels in Intestine 407 cells [8] and KB cells [7]. Also, outward rectification of heart-derived ClC3 channel (40 pS and 11 pS for outward and inward currents, respectively [19]) is much marked than that of VSOR Cl^- channel in Intestine 407 cells (37 pS at +50 mV and 27 pS at −50 mV [13]). ClC3 cloned from kidney cell lines also exhibited Cl^- current properties distinct from those of VSOR Cl^- current in a variety of cell types (see [4]), when expressed in *Xenopus* oocytes [20] or CHO cells [21]. For example, the Cl^- currents were inhibited by activation of protein kinase C or by an increase in cytosolic Ca^{2+}. Also, kidney-derived ClC3 exhibited much marked outward rectification and much larger single-channel conductance (100 or 140 pS) than VSOR Cl^- channels in many cell types [4]. However, there remains a possibility that ClC3 forms heteromeric channels with other ClC members. Unfortunately no study has been made by abolishing endogenous ClC3 expression. Before a precise conclusion, further studies are needed by carefully testing each *criterion*.

VOLUME-SENSITIVE BUT NOT MECHANO-SENSITIVE

Cell swelling should induce deformations of cell membrane and the underlying cytoskeletal network. Also, cell membrane should be stretched at an initial phase of swelling before deformation or unfolding of membrane invaginations or folds. Actually, activation of stretch-activated Ca^{2+}-permeable cation channel currents was demonstrated by whole-cell patch-clamp in Intestine 407 cells [22]. Since mechano-sensitive anion channels have been found in *E-Coli* [23, 24], it has been frequently inferred that the VSOR Cl^- channel is mechanically activated by swelling-induced membrane stretch. Under the whole-cell configuration, positive pressure application to the pipette was found to evoke VSOR Cl^- currents in some cell types [25, 26]. Because this maneuver had induced cell inflation, it is possible that the channel activation was due to membrane deformations but not to membrane stretch. Under the cell-attached configuration, suction application to the pipette never induced VSOR Cl^- channel activity in human Intestine 407 cells [4, 27] or did induce Cl^- channel activity with different activation kinetics in bovine ciliary epithelial cells [28]. Also, it is noteworthy that VSOR Cl^- channel activity was not inhibited by microfilament-disrupting cytochalasin D [4] and microtubule-disrupting vincristine (Fig. 3) in human epidermoid KB cells, although mechano-gating may be abolished by disrupting the interaction with cytoskeletal elements [29]. Thus, it is likely that membrane stretch is not a causal factor for activation of VSOR Cl^- channel.

Based on the data of membrane capacitance recorded from Intestine 407 and KB cells, it has been shown that the cell membrane has much space for swelling up to 3- to 5-fold volume increases, during which the capacitance did not largely change, due to electrically accessible membrane invaginations or foldings [4]. Thus, it appears that osmotic cell swelling brought about unfolding of invaginated membranes without producing a significant increase in membrane tension.

Single VSOR Cl^- channel events could not be recorded on swollen Intestine 407 cells when giga-seal between the attached pipette and the cell surface had been formed before the cell was rendered swollen [13, 27]. Therefore, it is unlikely that VSOR Cl^- channels were activated by some mobile factors, drop in osmolality, or dilution of cytosolic constituents. Since VSOR Cl^- channel activation was consistently observed when the pipette was attached to an Intestine 407 cell which had been already swollen, it is likely that channel activation was induced somehow by volume expansion-induced membrane deformation. Simultaneous measurements of cell size and whole-cell currents showed that the VSOR Cl$^-$ current density proportionally increased with increases in the surface area estimated by cell contour morphology of KB cells [7] and Intestine 407 cells [4]. The volume sensitivity of this current was not largely affected by vincristine (Fig. 3) but markedly increased by intracellular application of cytochalasin D [4].

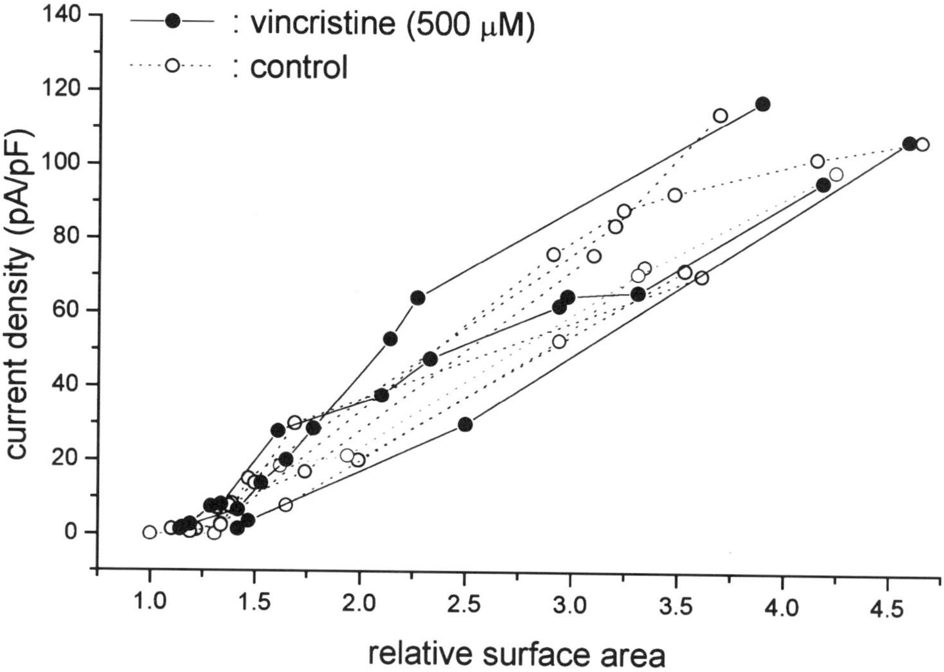

Figure 3. Volume sensitivity of whole-cell VSOR Cl$^-$ currents in osmotically swollen KB cells in the absence (open circles) and presence (filled circles) of 500 μM vincristine added to intracellular (pipette) solution. Single spherical KB-3-1 cells, which do not express PGP, were employed. (Methods: see [7]).

Taken together, it is concluded that activation of VSOR Cl$^-$ channel is not mediated by osmo-sensor or mechano-sensor but induced by volume expansion presumably due to unfolding of membrane invaginations or folds.

We thank K. Imoto and N. Sekiguchi for advice for RT-PCR study, M. Ohara for assistance with the illustrations, and M. Takazawa for secretarial help. This work was supported by a Grant-in-Aid for Scientific Research on Priority Areas of "Channel-Transporter Correlation" from the Ministry of Education, Science, and Culture of Japan, and by CREST (Core Research for Evolutional Science and Technology) of Japan Science and Technology Corporation (JST).

REFERENCES

1 Hazama A, Okada Y. Ca^{2+} sensitivity of volume-regulatory K^+ and Cl^- channels in cultured human epithelial cells. J Physiol 1988; 402: 687-702.

2 Hoffmann EK, Simonsen LO. Membrane mechanisms in volume and pH regulation in vertebrate cells. Physiol Rev 1989; 69: 315-382.

3 Okada Y, Hazama A. Volume-regulatory ion channels in a human epithelial cells. News Physiol Sci 1989; 4: 238-242.

4 Okada Y. Volume expansion-sensing outward-rectifier Cl^- channel: fresh start to the molecular identity and volume sensor. Am J Physiol 1997; 273: C755-C789.

5 Valverde MA, Diaz M, Sepúlveda FV. Volume-regulated chloride channels associated with the human multidrug-resistance P-glycoprotein. Nature 1992; 355: 830-833.

6 Gill DR, Hyde SC, Higgins CF, Valverde MA, Mintenig GM, Sepulveda FV. Separation of drug transport and chloride channel functions of the human multidrug resistance P-glycoprotein. Cell 1992; 71: 23-32.

7 Miwa A, Ueda K, Okada Y. Protein kinase C-independent correlation between P-glycoprotein expression and volume sensitivity of Cl^- channel. J Membrane Biol 1997; 157: 63-69.

8 Tominaga M, Tominaga T, Miwa A, Okada Y. Volume-sensitive chloride channel activity does not depend on endogenous P-glycoprotein. J Biol Chem 1995; 270: 27887-27893.

9 Vanoye CG, Altenberg GA, Reuss L. P-glycoprotein is not a swelling-activated Cl^- channels; possible role as a Cl^- channel regulator. J Physiol 1997; 502: 249-258.

10 Han ES, Altenberg GA, Reuss L. Swelling-activated Cl^- current associated with P-glycoprotein (Pgp) is inhibited by monoclonal antibody C219. J Gen Physiol 1994; 104: 43a (Abstract).

11 Paulmichl M, Li Y, Wickman K, Ackerman M, Peralta E, Clapham D. New mammalian chloride channel identified by expression cloning. Nature 1992; 356: 238-241.

12 Voets T, Buyse G, Tytgat J, Droogmans G, Eggermont J, Nilius B. The chloride current induced by expression of the protein pI_{cln} in Xenopus oocytes differs from the endogenous volume-sensitive chloride current. J Physiol 1996; 495: 441-447.

13 Tsumura T, Oiki S, Ueda S, Okuma M, Okada Y. Sensitivity of volume-sensitive

Cl^- conductance in human epithelial cells to extracellular nucleotides. Am J Physiol 1996; 271: C1872-C1878.
14 Gschwentner M, Nagl UO, Woll E, Schmarda A, Ritter M, Paulmichl M. Antisense oligonucleotides suppress cell volume-induced activation of chloride channels. Pflügers Arch 1995; 430: 464-470.
15 Gründer S, Thiemann A, Pusch M, Jentsch TJ. Regions involved in the opening of ClC-2 chloride channel by voltage and cell volume. Nature 1992; 360: 759-762.
16 Thiemann A, Grunder S, Pusch M, Jentsch TJ. A chloride channel widely expressed in epithelial and non epithelial cells. Nature 1992; 356: 57-60.
17 Malinowska DH, Kupert EY, Bahinski A, Sherry AM, Cuppoletti J. Cloning, functional expression, and characterization of a PKA-activated gastric Cl^- channel. Am J Physiol 1995; 268: C191-C200.
18 Lorenz C, Pusch M, Jentsch TJ. Heteromultimeric ClC chloride channels with novel properties. Proc Natl Acad Sci USA 1996; 93: 13362-13366.
19 Duan D, Winter C, Cowley S, Hume JR, Horowitz B. Molecular identification of a volume-regulated chloride channel. Nature 1997; 390: 417-421.
20 Kawasaki M, Uchida S, Monkawa T, Miyawaki A, Mikoshiba K, Marumo F, Sasaki S. Cloning and expression of a protein kinase C-regulated chloride channel abundantly expressed in rat brain neuronal cells. Neuron 1994; 12: 597-604.
21 Kawasaki M, Suzuki M, Uchida S, Sasaki S, Marumo F. Stable and functional expression of the ClC-3 chloride channel in somatic cell lines. Neuron 1995; 14: 1285-1291.
22 Okada Y, Hazama A, Yuan W-L. Stretch-induced activation of Ca^{2+}-permeable ion channels is involved in the volume regulation of hypotonically swollen epithelial cells. Neurosci Res 1990; 12: S5-S13.
23 Martinac B, Adler J, Kung C. Mechanosensitive ion channels of E. coli activated by amphipaths. Nature 1993; 348: 261-263.
24 Sukharev SI, Blount P, Bartinac B, Blattner FR, Kung C. A large-conductance mechanosensitive channel in E. coli encoded by mscL alone. Nature 1994; 368: 265-268.
25 Doroshenko P, Neher E. Volume-sensitive chloride conductance in bovine chromaffin cell membrane. J Physiol 1992; 449: 197-218.
26 Lewis RS, Ross PE, Cahalan MD. Chloride channels activated by osmotic stress in T lymphocytes. J Gen Physiol 1993; 101: 801-826.
27 Okada Y, Petersen CCH, Kubo M, Morishima S, Tominaga M. Osmotic swelling activates intermediate-conductance Cl^- channels in human intestinal epithelial cells. Jpn J Physiol 1994; 44: 403-409.
28 Zhang JJ, Jacob TJC. Three different Cl^- channels in the bovine ciliary epithelium activated by hypotonic stress. J Physiol 1997; 499: 379-389.
29 Hamil OP, McBride DW Jr. The cloning of a mechano-gated membrane ion channel. TINS 1994; 17: 439-443.

1998 Elsevier Science B.V.
Cell Volume Regulation: The Molecular Mechanism and Volume Sensing Machinery
Y. Okada, editor.

Platelet endothelial cell adhesion molecule-1 (PECAM-1, CD31) is involved in mechano-signal transduction in endothelial cells.

M. Masuda, M. Osawa, H. Shigematsu[¶], N. Harada, R. B. Lopes, K. Kusano and K. Fujiwara

Department of Structural Analysis, National Cardiovascular Center Research Institute, Suita, Osaka 565-8565, Japan
[¶]Laboratory for Chemistry, Institute for Life Science Research, Asahi Chemical Industry Co., Ltd., Fuji, Shizuoka 416, Japan

INTRODUCTION

Cells in multicellular organisms experience not only changes in the chemical environment but also those in the physical environment. Mechanical forces, such as stretch, compression and fluid shear stress, are always exerted onto cells and act as external signals for certain cells. Vascular endothelial cells (ECs) are unique cells in our body because they are constantly exposed to high levels of hemodynamic forces of flowing blood. The pulsatile flow in the blood vessel generates two primary forces acting on the vessel wall where endothelial cells exist (Fig. 1). One is a frictional force acting tangentially on the luminal surface and the force per unit area is called fluid shear stress. The second one is the transmural pressure and its changes cause diameter changes of the visco-elastic vessel wall and thus generate stretch force on the wall cells. A number of *in vivo* and *in vitro* evidence has been accumulated that both of these flow-related forces

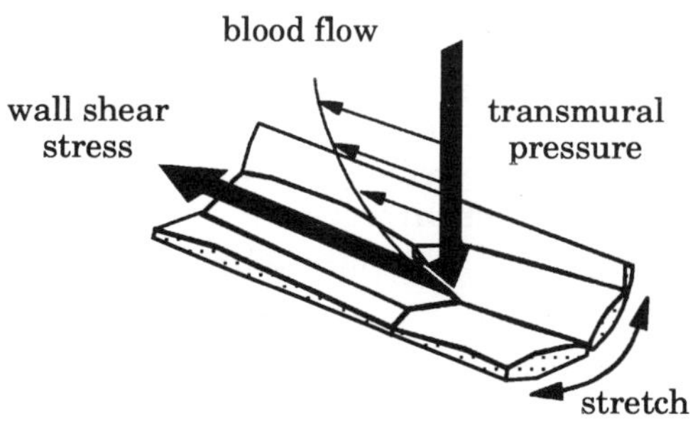

Figure 1. Hemodynamic forces acting on the blood vessel wall. The wall shear stress is the product of the velocity gradient of the fluid flow (fluid shear rate) at the wall surface and the viscosity of the fluid. If the vessel is a straight tube and the wall surface is smooth, the wall shear stress ($\tau_{w,}$, dyne/cm^2) is given as,

$$\tau_w=4Q\cdot\mu/\pi\cdot r^3,$$

where Q is the flow rate (ml/sec), μ is the viscosity (P=dyne·sec/cm^2) and r (cm) is the radius of the vessel. The physiological range of wall shear stress is estimated as 10-30 dynes/cm^2 in arteries and 1-6 dynes/cm^2 in veins. These values are 10^3-10^4 times smaller than those of the transmural pressure and 10^3 times smaller than those of the pressure difference caused by osmotic shock or suction through a patch pipette.

induce various responses in ECs including secretion of vasoactive substances, changes in gene expression and changes in the cytoskeleton and cellular morphology (see [1] for a review).

Although molecular mechanisms for flow sensing and subsequent signaling events in ECs are largely unknown, a transient Ca^{2+} mobilization [2, 3, 4, 5], which may be caused mainly by phosphoinositide (PI) turnover [6, 7], and increased K^{+}-efflux [8, 9], which may come from certain K^{+}-channel activation [8], are characterized as early events induced by fluid shear stress (or fluid shear rate). Involvement of gadolinium-sensitive cation channels (stretch-activated cation channels) in mechano-response has been shown in ECs subjected to mechanical stretch [10, 11]. However, these channels may not be essential for flow-induced morphological changes in ECs [12].

When the levels of fluid shear stress that induce various EC responses are compared, there are at least two or three different threshold levels, each for a different type of cell response. Ca^{2+} mobilization and K^{+}-efflux occurs when ECs are exposed to 1 dyne/cm^2 or less of shear stress [2, 3, 4, 5, 8] whereas many of the flow-induced gene expression and morphological changes appear to be initiated by 4 to 10 dynes/cm^2 of shear stress [1]. Thus, there may be multiple sensing mechanisms for fluid flow in ECs. It is possible that the EC responses with low threshold levels, such as the Ca^{2+} transient, are mainly attributed to the shear rate-dependent change in the agonist delivery rate onto the cell surface [3, 4, 13]. Some other responses initiated only by higher levels of shear stress may be the results of mechanical effects of flow, *i. e.* shear stress *per se*. Although the absolute magnitude of shear stress is much smaller than that of the normal pressure (see Fig. 1 legend), 10 dynes/cm^2 of shear stress can induce cellular deformation [14], which may trigger mechano-responses in ECs. Actually, we as well as others have shown that some EC responses are dependent on shear stress but not on shear rate [15, 16, 17]. Slow and sustained phase of the Ca^{2+} mobilization may also be the result of shear stress induced Ca^{2+}-influx [13, 18]. However, the molecular aspects of the mechano-sensing and subsequent signal transducing events in ECs have not been elucidated. Recently we have found that platelet endothelial cell adhesion molecule 1 (PECAM-1) is rapidly tyrosine phosphorylated in ECs exposed to physiological levels of flow [19, 20]. The known characteristics of this molecular event described below suggest that PECAM-1 tyrosine phosphorylation is caused directly by mechanical perturbation of ECs and mediates a signal toward MAP kinase activation.

MECHANO-SENSITIVE TYROSINE PHOSPHORYLATION OF PECAM-1

To get a handle on some early molecular events in flow sensing by ECs, we surveyed changes in phosphotyrosine levels of cell surface proteins of cultured bovine aortic endothelial cells (BAECs) which were exposed to fluid flow. We found that a 128 kDa glycoprotein was tyrosine phosphorylated by shear stress levels above 4 dynes/cm^2 [19]. The phosphorylation was rapid, reached the maximum by 5 min, and sustained at least for 10 min (Fig. 2). We also found

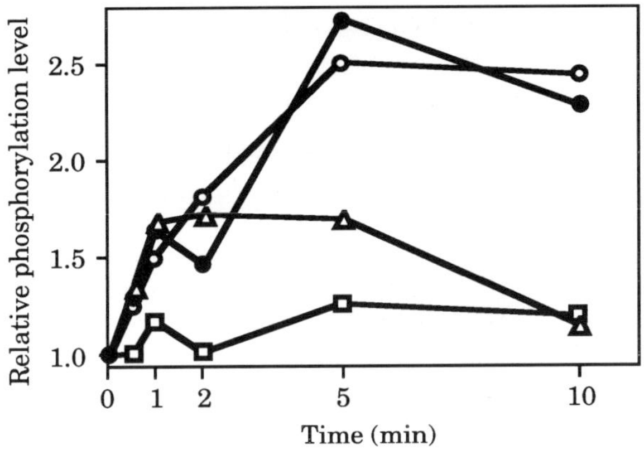

Figure 2. Time course of mechanical stress-induced PECAM-1 tyrosine phosphorylation. Confluent monolayers of BAECs were subjected to fluid shear stress (16 dynes/cm^2) using a cone-plate type flow apparatus (closed circles) or exposed to a hypertonic medium (open circles, a double osmolarity with sucrose), to a hypotonic medium (open triangles, a half osmolarity with 10 mM Hepes buffer), or to the normal culture medium (open squares). Relative tyrosine phosphorylation levels of 128 kDa glycoprotein (PECAM-1) against time 0 control cells were measured by Western blot analyses.

that the similar tyrosine phosphorylation was induced by hyper-osmotic shock and a transient one was initiated by hypo-osmotic shock (Fig. 2). The 128 kDa glycoprotein was then identified as a bovine homologue of PECAM-1 by microsequence analyses of the purified glycoprotein and subsequent cloning of its cDNA [20]. Cultured ECs obtained from human umbilical cord arteries and veins also showed the similar mechano-sensitive PECAM-1 tyrosine phosphorylation [20].

The PECAM-1 tyrosine phosphorylation induced by flow or osmotic changes was not inhibited by Ba^{2+} or tetraethyl ammonium ion (Fig. 3), which are known as blockers of the putative flow-activated K^+-channels [8, 9], or by gadolinium ion, a stretch activated cation channel blocker. The tyrosine phosphorylation was not induced by thrombin, ATP (Fig. 3) or bradykinin [20]. These three agonists are known to be degraded on the EC surface and thus are the possible candidates for the external mediators that can induce the shear-rate dependent Ca^{2+} mobilization in ECs [3, 4, 13]. Forced Ca^{2+}-influx or PKC activation were rather inhibitory for the phosphorylation (Fig. 3). These results indicate that the PECAM-1 tyrosine phosphorylation is not a downstream event of K^+-channel activation or Ca^{2+} metabolism in ECs and suggest that there is a novel, mechano-sensitive signaling pathway involving protein tyrosine phosphorylation in these cells.

PECAM-1, also known as CD31 or endoCAM, is a 130 kDa plasma membrane glycoprotein and is expressed on platelets, ECs, monocytes, neutrophils and certain subsets of T-cells [21, 22, for a review see 23]. PECAM-1 is a member of

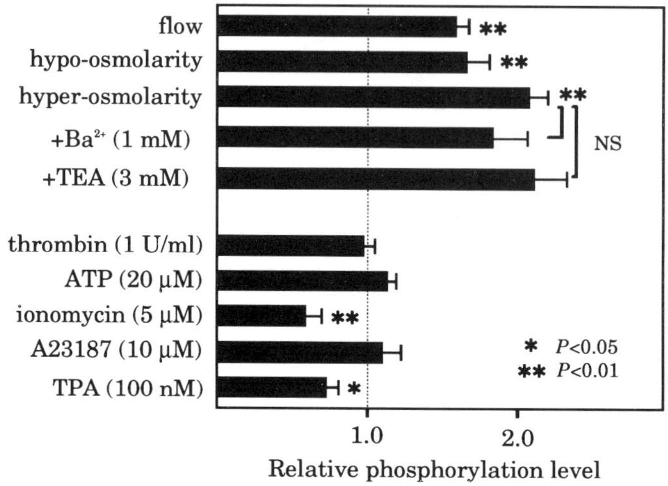

Figure 3. Effect of K^+-channel inhibitors and Ca^{2+} mobilizing reagents on PECAM-1 tyrosine phosphorylation. BAECs were exposed to flow (16 dynes/cm^2 for 5 min), osmotic shocks, Ca^{2+} mobilizing reagents, or a PKC activator TPA (5 min). The exposure time was 2 min except for flow and TPA stimulation. Results on hyper-osmotic shock in the presence of K^+-channel inhibitors are also shown.

the immunoglobulin (Ig) superfamily and has a large extracellular domain containing six Ig homology units, a single membrane spanning sequence, and a cytoplasmic tail. It is involved in cell-cell adhesion/recognition by a homophilic-binding mechanism [24] and, though there is a controversy [23], also by heterophilic binding to heparan sulfate proteoglycan [25] or other ligands such as $\alpha_v\beta_3$ integrin [26]. Crucial roles of this molecule in the transendothelial migration of monocytes and neutrophils were demonstrated both *in vitro* and *in vivo* [27, 28]. In ECs, PECAM-1 is localized at cell-cell adhesion sites (see Fig. 5) and participates in the inter-endothelial cell adhesion [22]. A rapid tyrosine phosphorylation of PECAM-1 has been reported recently during platelet aggregation induced by thrombin [29]. However, PECAM-1 in thrombin-activated but non-aggregated platelets was not phosphorylated on tyrosine residues [29], suggesting again a coupling between mechanical states of cells and PECAM-1 tyrosine phosphorylation.

POSSIBLE MECHANISMS FOR THE MECHANO-SENSITIVE PECAM-1 TYROSINE PHOSPHORYLATION

To get insights for the molecular events that are upstream to the PECAM-1 tyrosine phosphorylation in mechanically stimulated ECs, we examined whether autocrine and/or paracrine mechanisms were involved in the phosphorylation. An isotonic mixture of hyper- and hypo-osmotic media which had been used to osmotically stimulate ECs did not have the activity to induce the PECAM-1 phosphorylation in non-stimulated ECs. It is possible that the phosphorylation is mediated by mediators whose life time is very short. One of such mediator is nitric oxide (NO) and NO secretion from ECs is activated by fluid flow [1]. On the other hand, the concentration of a short-lived, membrane permeable mediator such as NO in and around the synthesizing cells may be affected by mass transport accompanied by bulk water movement. However, neither NO donors (NOR-1 and SNAP) nor NO synthase inhibitors (L-NNA and L-NAME) affected tyrosine phosphorylation levels of PECAM-1 of non-treated ECs or those stimulated by hyper-osmotic shock. Finally, some non-specific damage of the plasma membrane by mechanical stress might cause a leak of membrane impermeable mediators, especially ligands for receptor tyrosine kinases from ECs. However, exogenous FGF or VEGF at a high dose (100 ng/ml) did not significantly induce the phosphorylation. These results suggest that the PECAM-1 tyrosine phosphorylation is not regulated by autocrine factors whose cell surface concentration is changed by cellular responses to mechanical stimuli or by mass transport mechanism.

The cell-cell junction, where PECAM-1 is localized, is the site where actin-based cytoskeletal structures are anchored to the plasma membrane and therefore is one of the potential sites where the mechanical stress generated by external forces is focused. Actually, PECAM-1 is colocalized with catenins, cadherin associated proteins connecting cadherin to actin filaments, in cultured ECs in a confluent monolayer (Fig. 5, A and B). PECAM-1 was also reported to be coimmunoprecipitatable with β-catenin [30]. Although there is no evidence that shows direct interaction between PECAM-1 and catenins, there may be indirect or weak one by which PECAM-1 is linked to the actin cytoskeleton. Interestingly, PECAM-1 tyrosine phosphorylation induced by osmotic shocks was almost abolished in ECs pretreated by an actin filament disrupting agent cytochalasin D and was potentiated by phalloidin pretreatment which stabilized actin filaments and led hypertrophy of the actin cytoskeleton in cells (Fig. 4). In

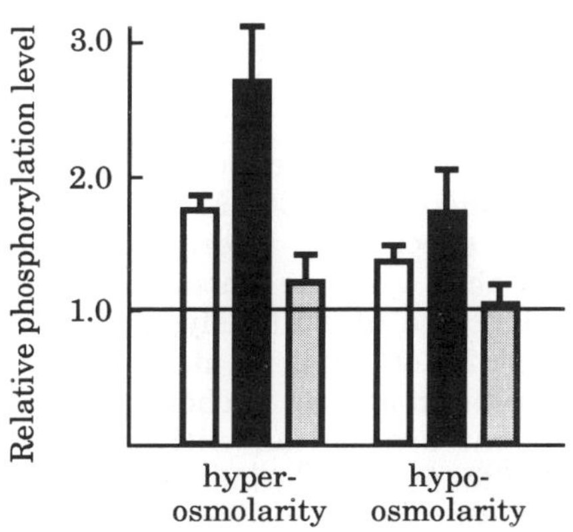

Figure 4. Integrity of the actin cytoskeleton affects PECAM-1 tyrosine phosphorylation induced by osmotic shocks. BAECs were incubated with 0.2% DMSO for 6 hr (open bars; carrier control), pretreated with 10 μM phalloidin for 6 hr (black bars), or pretreated with 1 μM cytochalasin D for 15 min (hatched bars) and osmotically stimulated for 5 min in the presence of the reagents. Relative tyrosine phosphorylation ratios (mean and SEM) against non-stimulated cells are shown.

contrast, intervention of the microtubule cytoskeleton by Colcemid (disruption) or taxol (stabilization) did not affect the osmotic-shock induced PECAM-1 tyrosine phosphorylation. Therefore, the effects of actin cytoskeleton intervention may be a specific one and not caused by changes in general mechanical property of cells [31]. These results suggest important roles of the actin cytoskeleton in the PECAM-1 phosphorylation.

Although the molecular mechanism of PECAM-1 tyrosine phosphorylation in ECs or platelets has not yet been elucidated, our study suggests an involvement of Src family kinases. The PECAM-1 cytoplasmic domain does not have any protein kinase activity and therefore, unlike receptor tyrosine kinases, PECAM-1 should be a substrate of certain protein tyrosine kinases. Using immunoprecipitated PECAM-1 and GST-fusion proteins containing the PECAM-1 cytoplasmic domain, we found that c-Src phosphorylated PECAM-1 at Tyr686 (Tyr712 when the signal peptide is included) in the cytoplasmic domain *in vitro* [20]. We also found that Tyr686 was phosphorylated in ECs. Src family kinases are known to be accumulated in the cell-cell adherens junction [32] and to be activated in mechanically stimulated cells [33]. Our preliminary data showed a transient activation of c-Src in hyper-osmotically treated ECs. These results suggest that the activation of Src family kinases represents a common signal transducing event evoked by various mechanical stimuli. Quite recently, PECAM-1 was shown to be phosphorylated at two tyrosine residues (Tyr663 and Tyr686) in HEK293 cells transfected with human PECAM-1 [34]. This result is consistent with recent reports including ours that tyrosine phosphorylated PECAM-1 interacts with a protein tyrosine phosphatase SH-PTP2 in ECs [35] and platelets [29] (see below). However, our *in vitro* data suggested that c-Src did not phosphorylate Tyr663 [20]. There may be other different kinases for Tyr663, Src family kinases may phosphorylate both of the tyrosine residues *in vivo,* or there may be a certain kinase which phosphorylates both. In any case, the identification of the putative PECAM-1 kinase(s) and its activation pathway is of utmost importance for elucidating the molecular mechanisms of cellular responses to mechanical stimuli.

PECAM-1 MEDIATED MAP KINASE ACTIVATION IN MECHANICALLY STIMULATED ENDOTHELIAL CELLS

Protein tyrosine phosphorylation plays principal roles in intracellular signaling and regulates various cellular functions. To characterize the PECAM-1 mediated signaling event, it is crucial to identify proteins specifically interacting with tyrosine phosphorylated PECAM-1. We found that tyrosine phosphorylated PECAM-1 was a major binding protein to SH-PTP2, a ubiquitously expressed protein tyrosine phosphatase containing two SH2 domains, in mechanically stimulated ECs [35]. SH-PTP2 is a cytoplasmic protein and its anchoring to the plasma membrane by the SH2-domain-dependent association with tyrosine phosphorylated membrane proteins is a key step for its regulation (see below). A rapid translocation of SH-PTP2 into cell-cell adhesion sites, presumably by the

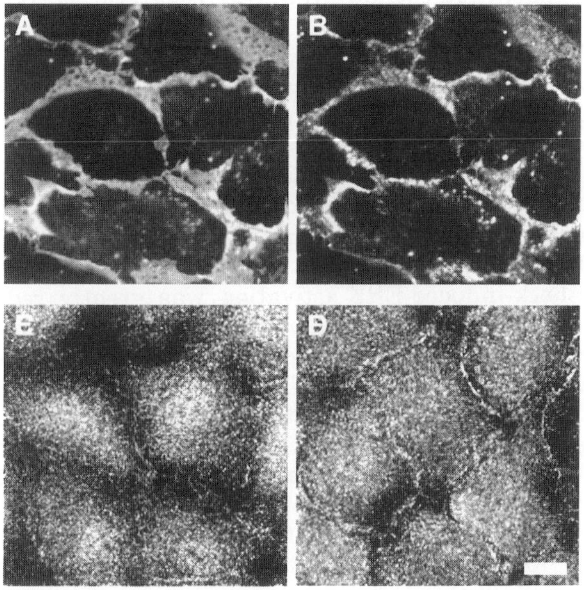

Figure 5. Localization of PECAM-1 in cell-cell adhesion sites and the SH-PTP2 translocation into these sites in ECs stimulated by hyper-osmotic shock. Non-stimulated HUVECs were doubly stained with anti-PECAM-1 (A) and anti-α-catenin (B). Non-stimulated HUVECs (C) and those stimulated by hyper-osmotic shock for 4 min (D) were stained with anti-SH-PTP2. Confocal scanning microscope images showing a single optical section (A and B) and extended focus images reconstructed from series of optical sections covering the whole cell height (C and D). Scale bar, 10 μm.

association with tyrosine phosphorylated PECAM-1, occurred in hyper-osmotically stimulated ECs (Fig. 5, C and D).

SH-PTP2, also known as PTP1D, PTP2C, Syp, and SHP2, was identified as a mammalian homologue of *Drosophila* tyrosine phosphatase corkscrew [36]. Corkscrew (csw) acts in concert with *Drosophila*-raf to positively transduce the signal from a receptor tyrosine kinase torso to the MAP kinase activation pathway [37]. Similarly, SH-PTP2 is known to act as a positive signaling molecule just downstream of receptor tyrosine kinases and to activate the MAP kinase (ERK) cascade in mammalian cells [38-41, see 42 for a review]. SH-PTP2 has two SH2 domains in the N-terminal half and the association of these SH2 domains with phosphotyrosine residues activates its phosphatase activity [43], which may be required for its signal transduction [38, 39]. Sequence motif in tyrosine phosphorylated proteins favorable for binding to the SH2 domains of SH-PTP2 [44], we have proposed that the N-terminal SH2 domain binds to phosphorylated Tyr663 (pTyr663) and the C-terminal one binds to pTyr686 [35].

Jackson et al. [34] have subsequently confirmed this and shown that the PECAM-1/SH-PTP2 association stimulates the catalytic activity of the phosphatase.

The association of SH-PTP2 to PECAM-1 suggests that PECAM-1 tyrosine phosphorylation mediates a positive signal for activating ERK in mechanically stimulated ECs. In ECs exposed to fluid flow, a transient activation of ERK, which depends on tyrosine kinase activity, has been reported [45, 46]. We found a similar ERK activation, peaked at 10 min and returned to the basal level by 30 min, in ECs stimulated by hyper-osmotic shock. The observed ERK activation is, at least partly, responsible for a number of mechano-sensitive gene expression induced by fluid flow [1] and osmotic shock.

The level of osmotic-shock-induced ERK activation was greatly reduced when PECAM-1 expression in ECs was downregulated by antisense S-oligonucleotides. Moreover, ERK activation was blocked by tyrosine phosphatase inhibitors and was abolished by a MEK inhibitor PD 98059. This tyrosine phosphatase activity dependent signaling might be peculiar to the osmotic-shock induced ERK activation, because the ERK activation induced by VEGF was not affected by tyrosine phosphatase inhibitors and because the level of PECAM-1 tyrosine phosphorylation itself was increased by these inhibitors. These results indicate that PECAM-1 is an essential component of a mechano-sensing/signal-transduction machinery in ECs and suggest that the tyrosine phosphorylation

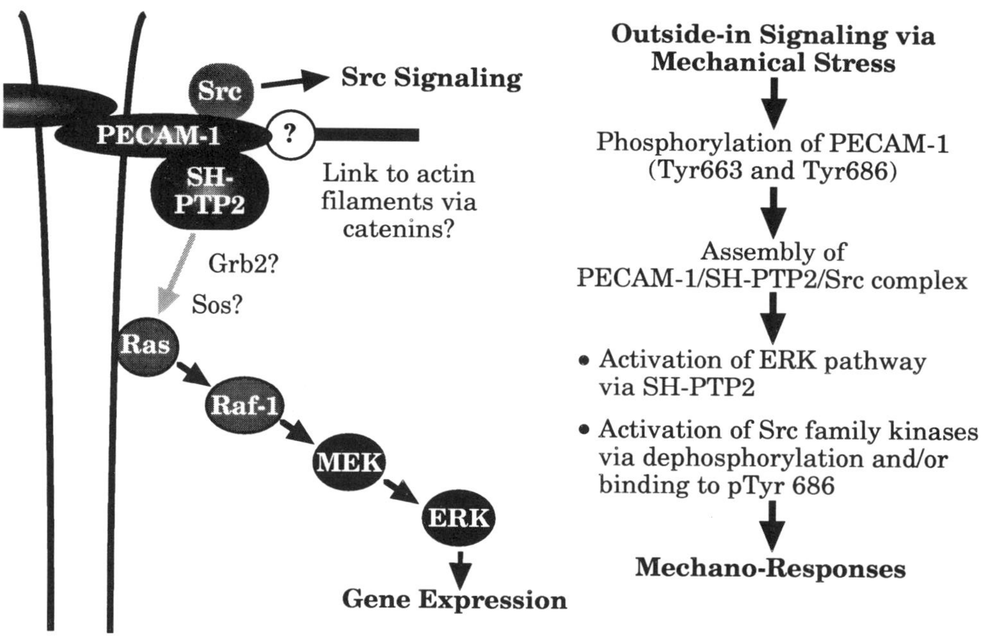

Figure 6. A model for transmembrane signaling of mechanical stimuli through PECAM-1.

dependent PECAM-1/SH-PTP2 interaction plays a major role in the mechano-signal transduction to the MEK-ERK pathway.

Another role of PECAM-1 in the mechano-signal transduction was also postulated [20, 34, 35]. As described earlier, c-Src is activated in mechanically stimulated ECs and may phosphorylates PECAM-1 at Tyr686. In addition, we found SH2 domain mediated c-Src binding to PECAM-1 at pTyr686 [20, 35]. Src family kinases are known to be inactivated by the intramolecular binding between their N-terminal SH2 domain and a phosphotyrosine residue in the C-terminal tail [48]. They can be activated either by the dephosphorylation of the C-terminal phosphotyrosine residue or by the competitive intermolecular interaction of the SH2 domain with high affinity binding motifs for the domain expressed on other proteins [48]. SH-PTP2 was shown to dephosphorylate the C-terminal phosphotyrosine residue [49] and tyrosine phosphorylated PECAM-1 provided a binding site for the Src family kinase SH2 domain while it did not interact with SH2 domains of PLCγ or PI3 kinase [35]. It is feasible that once PECAM-1 is tyrosine phosphorylated, it may effectively activate the nearby Src family kinases. If Src family kinases are actually involved in the PECAM-1 phosphorylation in mechanically stimulated ECs, then the Src/PECAM-1/SH-PTP2 interaction may work as a self-accelerating system for Src family kinases.

SUMMARY

The growing data described here suggest that a mechano-sensing machinery in which PECAM-1 is involved, is present at the interendothelial junction. When ECs are mechanically stimulated, such as by fluid flow and osmotic shock, PECAM-1 is tyrosine phosphorylated and provides a plasma membrane anchoring site for SH-PTP2 and possibly for Src family kinases. These interactions may effectively transmit an active signal to ERK and also amplify signaling mediated by Src family kinases (Fig. 6). In the junction, PECAM-1 binds to PECAM-1 expressed on the apposing EC and may also link to the actin cytoskeleton in the cytoplasm. Analogous to integrin mediated signaling [see 50 for a review], which may include mechano-sensing for cell adhesion onto substrates [51], the position of PECAM-1 in the molecular assembly in the junctional complex is favorable for the idea that PECAM-1 acts as one of the putative mechano-sensing molecules. The limited expression of PECAM-1 in ECs and some other blood cells and relative abundance of PECAM-1 in ECs (10^6 molecules/cell) [52] have led us to propose a hypothesis that PECAM-1 is involved in mechano-sensing and provides a molecular basis for the hyper-sensitivity of ECs to fluid flow stimulation.

Acknowledgments

This work was partly supported by a research grants from the Ministry of Health and Welfare of Japan (8B-1, 9A-4), Grants-in-Aid for Scientific Research from the Japanese Ministry of Education, and Special Coordination Funds for Promoting Science and Technology from Science and Technology Agency of Japan.

REFERENCES

1 Davies PF. Physiol Rev 1995; 75: 519-560.
2 Ando J, Komatsuda T, Kamiya A. In Vitro Cell Develop Biol 1988; 24: 871-877.
3 Dull R, Davies PF. Am J Physiol 1991; 261: H149-H154.
4 Mo M, Eskin SG, Schilling WP. Am J Physiol 1991; 260: H1698-H1707.
5 Shen J, Luscinskas FW, Connolly A, Dewey CF, Jr, Gimbrone MA, Jr. Am J Physiol 1992; 262: C384-C390.
6 Nollert MU, Eskin SG, McIntire LV. Biochem Biophys Res Comm 1990; 170: 281-287.
7 Prasad, ARS, Logan SA, Nerem RM, Schwartz CJ, Sprague EA. Circ Res 1993; 72: 827-836.
8 Olesen, AP, Calpham DE, Davies PF. Nature 1988; 331: 168-170.
9 Schilling, WP, Mo M, Eskin SG. Exp Cell Res 1992; 198: 31-35.
10 Naruse K, Sokabe M. Am J Physiol 1993; 264: C1037-C1044.
11 Wung BS, Cheng JJ, Chao YJ, Lin J, Shyy YJ, Wang DL. Am J Physiol 1996; 270: H1462-H1468.
12 Malek AM, Izumo S. J Cell Sci 1996; 109: 713-726.
13 Wiesner TF, Berk BC, Nerem RM. Proc Natl Acad Sci USA 1997; 94: 3726-3731.
14 Liu SQ, Yen M, Fung YC. Proc Natl Acad Sci USA 1994; 91: 8782-8786.
15 Masuda M, Fujiwara K. Front Med Biol Eng 1993; 5: 79-87.
16 Ando J, Ohtuka A, Korenaga R, Kawamura T, Kamiya A. Biochem Biophys Res Comm 1993; 190: 716-723.
17 Koller A, Sun D, Kaley G. Circ Res 1993; 72: 1276-1284.
18 Schwartz G, Droogmans G, Nilius B. Pflügers Arch 1992; 421: 394-396.
19 Harada N, Masuda M, Fujiwara K. Biochem Biophys Res Comm 1995; 214: 69-74.
20 Osawa M. Masuda M, Harada N, Lopes RB, Fujiwara K. Eur J Cell Biol 1997; 72: 229-237.
21 Newman, PJ, Berndt MC, Gorski J, White GC, II, et al. Science 1990; 247: 1219-1222.
22 Albelda SM, Oliver PD, Romer LH, Buck CA. J Cell Biol 1990; 110: 1227-1237.
23 Newman PJ. J Clin Invest 1997; 99: 3-8.
24 Fawcett, J, Buckley C, Holness CL, Bird IN, et al. J Cell Biol 1995; 128: 1229-1241.
25 Muller, WA, Berman ME, Newman PJ, DeLisser HM, Albelda SM. J Exp Med 1992; 175: 1401-1404.
26 Piali, L, Hammel P, Uherek C, Bachmann F, et al. J Cell Biol 1995; 130: 451-460.
27 Muller, WA, Weigl SA, Deng X, Phillips DM. J Exp Med 1993; 178: 449-460.
28 Vaporciyan, AA, DeLisser HM, Yan HC, Mendiguren II, et al. Science 1993; 262: 1580-1582.
29 Jackson DE, Ward CM, Wang R, Newman PJ. J Biol Chem 1997; 272: 6986-6993.
30 Matsumura T, Wolff K, Petzelbauer P. J Immunol 1997;158: 3408-3416.
31 Wang N, Butler JP, Ingber DE. Science 1993; 260: 1124-1127 .

32 Tsukita S, Oishi K, Akiyama T, Yamanashi Y. et al. J Cell Biol 1991; 113: 867-879.
33 Liu M, Qin Y, Liu J, Tanswell AK, Post M. J Biol Chem 1996; 271: 7066-7071.
34 Jackson DE, Kupcho KR, Newman PJ. J Biol Chem 1997; 272: 24868-24875.
35 Masuda M, Osawa M, Shigematsu H, Harada N, et al. FEBS Lett 1997; 408: 331-336.
36 Freeman RM Jr, Plutzky J, Neel BG. Proc Natl Acad Sci USA 1992; 89: 11239-11243.
37 Perkins LA, Larsen I, Perrimon N. Cell 1992; 70: 225-236.
38 Milarski KL, Saltiel AR. J Biol Chem 1994; 269: 21239-21243.
39 Noguchi T, Matozaki T, Horita K, Fujioka Y, Kasuga M. Mol Cell Biol 1994; 14: 6674-6682.
40 Xiao S, Rose DW, Sasaoka T, Maegawa H, et al. J Biol Chem 1994; 269: 21244-21248.
41 Roche S, McGlade J, Jones M, Gish GD, Pawson T, Courtneidge SA. EMBO J 1996; 15: 4940-4948.
42 Streuli M. Curr Opin Cell Biol 1996; 8: 182-188.
43 Pluskey S, Wandless TJ, Walsh CT, Shoelson SE. J Biol Chem 1995; 270: 2897-2900.
44 Case RD, Piccione E, Wolf G, Benett AM, et al. J Biol Chem 1994; 269:10467-10474.
45 Ishida T, Peterson TE, Kovach NL, Berk BC. Circ Res 1996; 79: 310-316.
46 Takahashi M, Berk BC. J Clin Invest 1996; 98: 2623-2631.
47 Jo H, Sipos K, Go YM, Law R, et al. J Biol Chem 1997; 272: 1395-1401.
48 Pawson T. Nature 1995; 373: 573-580.
49 Peng ZY, Cartwright CA. Oncogene 1995; 11: 1955-1962.
50 Dedhar S, Hannigan GE. Curr Opin Cell Biol 1996; 8: 657-669.
51 Pelham RJ Jr, Wang Y-l. Proc Natl Acad Sci USA 1997; 94: 13661-13665.
52 Newman PJ. Ann New York Acad Sci 1994; 714: 165-174.

Cell Volume Regulation: The Molecular Mechanism and Volume Sensing Machinery
Y. Okada, editor.

Neural volume regulation of single guinea-pig cardiac myocytes

T. Mitsuiye and A. Noma

Department of Physiology, Faculty of Medicine, Kyoto University, Kyoto 606, Japan.

INTRODUCTION

Ischemia of the cardiac tissue causes swelling of myocytes [1, 2, 3, 4] through depression of the Na^+-K^+ pump and accumulation of anaerobic metabolites. The cell swelling is in the vicious cycle since it compresses the vascular bed, and results in a decrease of compliance of the chamber. Although a local increase of catecholamines in the ischemic tissues is known [5, 6], its physiological relevance in relation to the cell swelling is unclear.

The membrane conductance due to the Cl^- channel activated by the stimulation of β-adrenergic receptor ($I_{Cl,cAMP}$) was first reported by Eagan et al [7] in 1988, and its ionic nature (Cl^- channel) was carefully examined by Matsuoka et al [8]. The cAMP-dependent pathway for the channel activation was also reported [9,10]. The molecular basis of $I_{Cl,cAMP}$ channel was revealed by Hume's group to be a membrane protein of the ATP-binding cassette (ABC) superfamily [11]. The $I_{Cl,cAMP}$ is very unique in respect that it appeares only under the sympathetic nervous control. However, its physiological function was still not clear since the membrane excitation is not dramatically changed by the $I_{Cl,cAMP}$ activation. We examined a possibility that the activation of $I_{Cl,cAMP}$ by β-adrenergic agonists may be responsible for a regulatory volume decrease (RVD), since an efflux of Cl^- in the presence of membrane K^+ conductance should be accompanied by a K^+ efflux, and finally causes osmosis of water to the cell exterior [4]. In fact, Wang et al., [12] found that isoprenaline induces RVD in the osmotically inflated ventricular myocytes through activation of $I_{Cl,cAMP}$ and this action is blocked by parasympathetic transmitter acetylcholine.

The other kind of Cl^- channels, which are responsible for the cell volume regulation, is the swelling induced Cl^- channels ($I_{Cl,swell}$) first described by Tseng [13] in 1992. The spontaneous RVD in the cultured cardiac myocytes was attributed to the activation of $I_{Cl,swell}$ by Zhang et al. [14]. It is reported that the characteristics of the $I_{Cl,swell}$ is different from $I_{Cl,cAMP}$ by Vandenberg et al. [15]. The $I_{Cl,swell}$ is activated by cell-inflation in the hypotonic solutions. However, it is still a matter of debate how extensively $I_{Cl,swell}$ is involved in the regulatory volume regulation in the hypotonic solutions. Usually RVD is hardly observed in the adult cardiac myocytes within 5-10 min of hypotonic stimulus [16] and the activation of $I_{Cl,swell}$ only occurred at a late stage of the hypotonic cell swelling [17]. This article briefly reviews the cardiac volume regulation through activation of $I_{Cl,cAMP}$ channels in guinea pig ventricular myocytes.

MEASUREMENTS OF CELL DIMENSION

It is essential to continuously measure the cell dimension to detect a slight change in the cell volume. The conventional method uses the video picture and the size of the cell image was measured. To improve the time resolution, we developed a new method to measure variations in the cell thickness of single ventricular cell by focusing a fine marker, which was a carbon particle and was attached on the upper surface of the ventricular myocytes [12]. The image of the particle was directed through the side-port of the microscope to an adjustable rectangular aperture and thence to the photomultiplier tube (photometerhead P100, Nikon). The size of the aperture was adjusted slightly smaller than the image of marker, thereby the output of the photomultiplier (light intensity) was minimum when the image of the particle was in focus. As the marker become out of focus by moving the objective glass up and down (± 5-10 μm) across the focus level, the light intensity increased almost exponentially. In the experiment, the objective glass was moved up and down by connecting the handle for fine adjustment of focus to an electric motor, which was driven by a sinusoidal signal every 6 sec. The relation between the focus position and the light intensity was displayed on the graphic display of the on-line computer. The first derivative of the light signal over the distance was almost linear near the minimum, and the focus position was determined as the intersection of the fitted line with the abscissa. The time course of the cell thickness changes was displayed on the graphic display during the course of the experiment.

The above method of measuring the cell thickness was performed using conventional apparatus for measuring Ca^{2+} signals, and can be combined potentially with the whole cell voltage clamp using the patch clamp technique. Automatic measurement of cell height was also achieved by focusing on fluorescent microbeads placed on both the apical and basal surfaces of cultured epithelial cells [18]. Similar but manual measurement of the thickness of ventricular myocytes was performed previously using small (2-3 µm) particles placed on the chamber surface and top of the cell [16]. The cell height was measured as the distance between these two markers. This method may be advantageous to our method, where the depth of the maker only on the top of the cell was measured. The measurement was therefore affected by a temperature-dependent change in the distance between the stage of the microscope and the objective glass (in the range of several µm). During experiment, it was necessary to keep the room temperature constant as much as possible. Also, the perfusion rate of the recording chamber with a pre-heated saline was kept constant. It is awaited to improve the method to decrease the temperature-dependent drift in the distance between the stage of the microscope and the objective glass to obtain a stable recording of the cell thickness over an experimental time of more than 5-10 min.

RVD INDUCED BY THE APPLICATION OF ISOPRENALINE

It has been reported that changes in the cell width and thickness are significantly greater than changes in cell length when dissociated cardiac myocytes swell in hypotonic solution [16,19]. The average thickness of guinea-pig ventricular myocyte was 7.8 ± 0.2 µm (n = 94) in the control Tyrode solution. In the control run of experiment, the relative cell thickness were 130.4 ± 3.1% (n=10), 119.1 ± 1.1% (n=50), 87.2 ± 1.9% (n=9) and 75.6 ± 3.2% (n=5) control in the 50%, 70%, 130% and 200% osmolarity, respectively. To examine any contribution of $I_{Cl,swell}$ in these measurements, effects of blocking $I_{Cl,swell}$ with 9AC were examined. In the presence of 500 µmol/L 9AC, the relative thickness were 129.2 ± 5.5% (n=6), 116.9 ± 1.8% (n=8), 84.0 ± 1.0% (n=7) and 77.8 ±1.4% (n=4) in 50%, 70%, 130% and 200% control osmolarity solutions, respectively. These values are in good agreement with the control run in the absence of the drug.

It is a consensus that the spontaneous RVD is minimal in the adult cardiac myocytes.

In the experiment shown in Fig. 1A, the cell thickness started to increase immediately after switching the external solution to the 70% hypotonic solution, and reached a new steady level within about 130 sec. The increase of thickness reversed only after switching back to the control Tyrode solution. We failed to observe a remarkable spontaneous RVD in the 70% and 50% hypotonic solutions within the experimental time of 1-15 min. Also no regulatory volume increase (RVI) was observed after the cell shrinkage in the 130% hypertonic solutions.

We examined if the activation of $I_{Cl,cAMP}$ can induce a volume decrease during the cell swelling in the hypotonic solutions. Since isoproterenol >50 μmol/L fully activates $I_{Cl,cAMP}$ [20], 50 nmol/L isoproterenol was applied after the cell thickness reached a new steady level in the 70% hypotonic solution (Fig. 1B). Indeed the cell thickness gradualy decreased after the application of isopreterenol and returned to the control level after about 10 min. This RVD induced by isoproterenol was observed in 11 of 18 cells (with 50 nmol/L isoproterenol). The rate of observing RVD was not increased by increasing isoprenaline concentration, 9 of 16 cells (with 0.1 μmol/L) and 5 of 11 cells (with 1 μ mol/L). In the non-responding cells, the cell thickness was not obviously changed, or the change in the thickness could not be resolved because of an artifactual drift of the measurement of unknown origin.

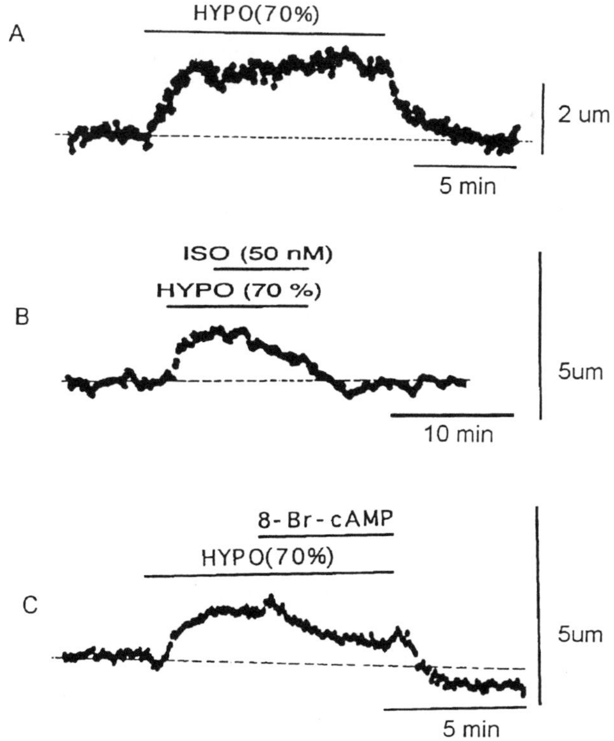

Fig. 1. Changes in the cell thickness of single ventricular cell of guinea pig. The perfusion of test solutions are indicated by the horizontal bars. The dotted lines indicate the basal thickness of the myocytes. The composition of the control Tyrode solution was (mmol/L) NaCl 140.0, KCl 5.4, $CaCl_2$ 1.8, $MgCl_2$ 0.5, NaH_2PO_4 0.3, glucose 5.5, and HEPES 5.0. The pH was adjusted to 7.4 with NaOH. The hypotonic solution of 70% or 50% control osmolarity was made by decreasing the concentration of NaCl to 100 or 70 mmol/L in the control Tyrode solution, respectively and the hypertonic solution of 130% or 200% control osmolarity was prepared by adding 90 or 280 mmol/L mannitol to the control Tyrode's solution. When the hypotonic solutions were applied, the control solution, prepared by adding mannitol to the hypotonic solutions was applied beforehand. In all experiments the temperature of the bath solution was kept at 35 ± 0.5℃ by passing the perfusion tube through a temperature-controlled water jacket prior to the inlet of the recording chamber.

The variation in the cell thickness was measured by focusing a small marker on the top surface of the myocyte. The lower surface of the myocyte was fixd on the glass bottom of the recording chamber mounted on an inverted microscope. To facilitate cell attachment, the bottom was cleaned with chromic acid and was coated with 100μg/L polylysine. After myocytes settled on the glass bottom, a suspension of small carbon particle (3-7 μm in diameter) was superfused. Single myocytes marked with a carbon particle near the middle of the cell was used. Sometimes we catched a proper carbon particle at the tip of a conventional glass pipette and positioned it to the cell surface if necessary. This system only detects a movement of the marker position in the vertical axis. The thickness of the ventricular myocyte was measured at the beginning of the experiment as the distance of the focus level of the markers on the top of the cell from the focus level of the bottom of the recording chamber.
Data are given as mean ± SEM. Statistical significance was determined using Student's unpaired *t* test.

The involvement of $I_{Cl,cAMP}$ activation in RVD was proved by various experimental findings. The application of isoprenaline or 8-Br-cAMP, both activators of $I_{Cl,cAMP}$, was able to induce RVD (Fig. 1C). Furthermore, the isoprenaline-induced RVD was blocked by 9AC and by ACh [8,9,10,20,21,22]. The former blocks the $I_{Cl,\ cAMP}$ channel directly, and the latter antagonizes the activation of adenylate cyclase mediated by the GTP-binding protein (Gi). The isoprenaline-induced RVD was observed in the presence of DIDS, which causes a rapid and complete inhibition of $I_{Cl,\ swell}$ [15]. Finally, the whole cell voltage clamp experiments confirmed the activation of $I_{Cl,\ cAMP}$ during RVD in the osmotically inflated myocytes.

MECHANISMS UNDERLYING RVD INDUCED BY β-ADRENERGIC STIMULATION

The isoprenaline-induced RVD can be explained as follows. The value of E_{Cl} in the cardiac myocyte ranges from -45 to -65 mV [23], therefore, at the resting membrane potential near to -90 mV, intracellular Cl^- is driven in the outward direction through the membrane. Thus, intracellular Cl^- should be extruded via activation of $I_{Cl,\ cAMP}$ channel and depolarize the membrane. Then, the membrane depolarization induces an extra efflux of cellular K^+, which is driven by the increased electromotive force largely through inward rectifier K^+ channels. The outward movement of K^+ and Cl^- results in an outflow of cellular water and thus in RVD. This view is consistent with the experimental finding of the undershoot of cell volume change on returning to the iso-osmotic solution after RVD was established (Fig. 1).

Based on the same rationale [14, 24], activation of $I_{Cl,swell}$ along with activation of K^+ and/or nonselective cation channels is considered to be responsible for the induction of the spontaneous RVD, which has been reported in many cell types [25]. However, the spontaneous RVD could not be observed with the experimental time of observation in the present study (over 10 min). This finding is consistent with the previous studies in rabbit ventricular cells [16]. This apparent contradition may be explained by the delayed activation of $I_{Cl,swell}$, which occured later than the activation of both the delayed K^+ current and the Na^+/K^+ pump in the 70% hyposmotic solution [17,26].

According to the above hypothesis, the activation of $I_{Cl,cAMP}$ in the normoosmotic solution should also induce a volume decrease of the ventricular myocytes. This should be examined in future experiments.

The isoprenaline-induced RVD during the whole cell voltage clamp is apparently inconsistent with the above hypothesis. Since the membrane potential was maintained constantly during activation of $I_{Cl,\ cAMP}$, the passive depolarization-driven extra-movement of K^+ or Cs^+ is not expected across the cell membrane. However, it is necessary to consider the ionic flux through the tip of the whole cell patch clamp electrode. The membrane Cl^- flux through the membrane should be balanced by the movement of both cations and anions through the electrode tip. The cations is extruded through the electrode tip when the inward $I_{Cl,cAMP}$ (efflux of Cl^-) was activated. These movements of cations through the electrode tip and the anions through the ion channels should result in a decrease in the ionic solute within the myocytes during the voltage clamp. This hypothesis may be examined by conducting the whole cell voltage clamp combined with the measurement of the cell volume.

It should be noted that the basal activity of $I_{Cl,cAMP}$ is none in the absence of the β-adrengergic agonists. We propose that the major physiological function of the $I_{Cl,cAMP}$ channel is to regulate the cell volume. The importance of myocardial cell volume regulation is evident from the fact that cell swelling reduces chamber compliance [27], and occludes capillaries of the coronary circulation. The release of catecholamines is increased in the ischemic and hypoxic tissues [5,6]. So, RVD induced through activation of $I_{Cl,cAMP}$ may operate to relieve the myocytes from this vicious cycle. Furthermoer, since the heart is continuously under the influence of both the adrenergic and muscarinic stimulation, it is possible that autonomic nerves regulate the cardiac cell volume in addition to the regulation of the membrane excitability.

ACKNOWLEDGMENTS:

The authors thank Drs. Z. Wang, N. Sasaki and Dr. M. Takano for their discussion. This study was supported by a scientific research grant from the Ministry of Education, Science and Culture of Japan.

REFERENCES:

1. Leaf, A. Cell swelling. *Circulation.* 1973;48(3):455-458
2. Jennings, RB. et al. Effect of reperfusion late in the phase of reversible ischemic injury. *Circ Res.* 1985;56:262-278.
3. Tranum-Jensen, J. et al. Tissue osmolality, cell swelling, and reperfusion in acute regional myocardial ischemia in the isolated porcine heart. *Circ Res.* 1981;49:364-381
4. MacKnight ADC, Leaf A. Regulation of cellular volume. *Physiol Rev.* 1977;57:510-573.
5. Videbaek, J., Christensen, NJ., Sterndorff, B. Serial determination of plasma catecholamines in myocardial infarction. *Circulation.* 1972;46:846-855.
6. Karawatowska-Krynska, E., Beresewicz, A. Effects of locally released catecholamines on lipolysis and injury of the hypoxicisolated rabbit heart. *Journal of Molecular and Cellular Cardiology.* 1983;15:523-536.
7. Egan, TM., Noble, D., Noble, SJ., Powell, T., Twist, VW., Yamaoka, K. On the mechanism of isoprenaline- and forskolin- induced depolarization of single guinea pig ventricular myocytes. *J Physiol (London).* 1988;400:299-320.
8. Matsuoka, S., Ehara, T., Noma, A. Chloride-sensitive nature of the adrenaline-induced current in guinea-pig cardiac myocytes. *J. Physiol (London).* 1990;425:579-598.
9. Bahinski, A., Nairin, AC., Greengard, P., Gadsby, DC. Chloride conductance regulated by cyclic AMP-dependent protein kinase in cardiac myocytes. *Nature.* 1989;340:718-721.
10. Harvey, RD., Hume, JR. Autonomic regulation of a chloride current in heart. *Science.* 1989;244:983-985.
11. Hart, P., J. D. Warth, P. C. Levesque, M. L. Collier, Y. Geary, B. Horowitz, and J. R. Hume. 1996. Cystic fibrosis gene encodes a cAMP-dependent chloride channel in heart. *Proc. Natl. Acad. Sci. USA.* 93:6343-6348.
12. Wang Z., Mitsuiye T., Rees S.A. and Noma A. Regulatory volume decrease of cardiac myocytes induced by β-adrenergic activation of the Cl^- channel in guinea pig. J. Gen. Physiol. 1997;110:73-82.
13. Tseng GN. Cell swelling increases membrane conductance of canine cardiac cells: evidence for a volume-sensitive Cl channel. *Am J Physiol.* 1992;262:C1056-C1068.
14. Zhang J, Rasmusson RL, Hall SH, Lieberman M. A chloride current associated with swelling of cultured chick heart cells. *J Physiol (London).* 1993;472:801-820.
15. Vandenberg JI, Yoshida A, Kirk K, Powell T. Swelling-activated and isoprenaline-activated chloride currents in guinea-pig cardiac myocytes have distinct electrophysiology and pharmacology. *Gen J Physiol.* 1994;104:997-1017.
16. Drewnowska, K., Baumgarten, CM. Regulation of cellular volume in rabbit ventricular myocytes: bumetanide, chlorothiazide, and ouabain. *Am J Physiol.* 1991;260:C122-C131.
17. Sasaki N, Mitsuiye T, Wang Z, Noma A. Increase of the delayed rectifier K^+ and Na^+-K^+ pump current by hypotonic solutions in guinea pig cardiac myocytes. *Circ Res.* 1994;75:887-895.
18. Van Driessche, W., P. D. Smet, and G. Raskin. 1993. An automatic monitoring system for epithelial cell height. *Pflüg. Arch. Eur. J. Physiol.* 425:164-171.
19. Roos, KP. Length, width, and volume changes in osmotically stressed myocytes.

Am. J Physiol. 1986;251:H1373-H1378.

20. Tareen FM, Ono K, Noma A, Ehara T. β-adrenergic and muscarinic regulation of the chloride current in guinea-pig ventricular cells. *J Physiol (London).* 1991; 440:225-241.
21. Harvey, RD., Clark, CD., Hume, JR. Chloride current in mammalian cardiac myocytes. Novel mechanism for autonomic regulationof action potential duration and restion membrane potential. *Gen J Physiol.* 1990;95:1077-1102.
22. Levesque, PC., Clark, CD., Zakarov, SI., Rosenshtraukh, LV., Hume, JR. Anion and cation modulation of the guinea-pig ventricular action potential during β-adrenoceptor stimulation. *Pflügers Arch.* 1993;424:54-62.
23. Vaughan-Jones RD. Chloride activity and its control in skeletal and cardiac muscle. *Philos Trans R Soc Lond [Biol]* 1982;299:537-548.
24. Hoffmann EK, Simonsen LO. Membrane mechanisms in volume and pH regulation in vertebrate cells. *Physiol Rev.* 1989;69:315-377.
25. Sarkadi B, Parker JC. Activation of ion transport pathways by changes in cell volume. *Biochimica et Biophysica Acta.* 1991;1071:407-427.
26. Wang, Z., Mitsuiye, T., Noma, A. Cell distension-induced increase of the delayed rectifier K^+ current in guinea pig ventricular myocytes. *Circ Res.* 1996;78:466-474.
27. Vogel WM, Cerel AW, Apstein CS. Post-ischemic cardiac chamber stiffness and coronary vasomotion: The role of edema and effects of dextran. *J Mol Cell Cardiol.* 1986;18:1207-1218.

Cell Volume Regulation: The Molecular Mechanism and Volume Sensing Machinery
Y. Okada, editor.

Regulatory volume decrease in cultured brain cells: rate limiting factors

Pasantes-Morales, H., Cardin, V. Morales-Mulia, S. and Quesada, O.

Institute of Cell Physiology, National University of Mexico, Mexico City, Mexico.

Brain cells in culture, astrocytes as well as neurons, when exposed to hyposmotic media first swell and then activate a process of cell volume recovery, consisting in a rapid initial phase occurring during the first 5-6 min after swelling followed by a slower phase lasting about 10 min during which cells recover their original volume with an efficiency depending upon the type of cell and the decrease in osmolarity (1,2-5). This regulatory volume decrease (RVD) is accomplished by extrusion of intracellular osmotically active solutes. The ions K^+ and Cl^- and a number of organic molecules, predominantly amino acids, serve this function as osmolytes (3-6). In cultured astrocytes the swelling activated efflux of these osmolytes occurs essentially by separate K^+ and Cl^- channels for ions and by a leak pathway for amino acids, which shares with the anion channel a number of physiological and pharmacological features (7-9). The molecular identity of these channels has not been conclusively established. For the anion channel, possible candidates include the band 3-like erythroid anion exchanger, the P-glycoprotein, the phospholemman, the I_{Cln} or other still non identified Cl^- channels different from those already cloned. Some of these molecules, as discussed in other chapters, may function as regulators of the anion channel itself (9,10). The volume sensitive K^+ channel has not been studied in such detail despite the fact that K^+ fluxes are the rate limiting step of RVD as discussed below.

In cultured astrocytes, the time course of release of the main osmolytes K^+, Cl^- and amino acids is markedly different. Whereas Cl^- efflux activates immediately after swelling and also rapidly inactivates in less than 80 sec, the efflux of taurine follows a time course more close to the change in cell volume, with a peak activation at about 1 min and inactivation starting within the next 2-3 min. In contrast, the efflux of K^+ exhibits a slower time course with maximal activation at about 2 min followed by a plateau of 2-5 min before initiating the inactivation phase (Fig. 1). Also differences are observed regarding the amount of osmolyte released. We cannot properly assess the net Cl^- loss since it moves very rapidly and we measure unidirectional fluxes using ^{125}I as tracer but the decrease in K^+ and amino acids can be more accurately determined. In cultured astrocytes, a large amount (30-80%) of the cell pool of amino acids acting as osmolytes is released in response to a 50% hyposmotic solution, but the proportion of K^+ released is much lower (about 21%). These two features of

K^+, i.e. the slow activation/inactivation and the amount released, may explain why K^+ efflux seems to be the rate limiting step of RVD in astrocytes. It has been consistently observed that increasing K efflux by different procedures increase the rate and efficiency of RVD. The most obvious is the opening of an alternate route for K^+ by the ionophores gramicidin and valinomycin but other conditions leading to marked acceleration of RVD due to an increase in K^+ efflux include removal of NaCl, omission of external Ca^{2+} and exposure to ATP. These effects have to be considered in the interpretation of results related, for instance, to the influence of Ca^{2+} on RVD, as discussed below.

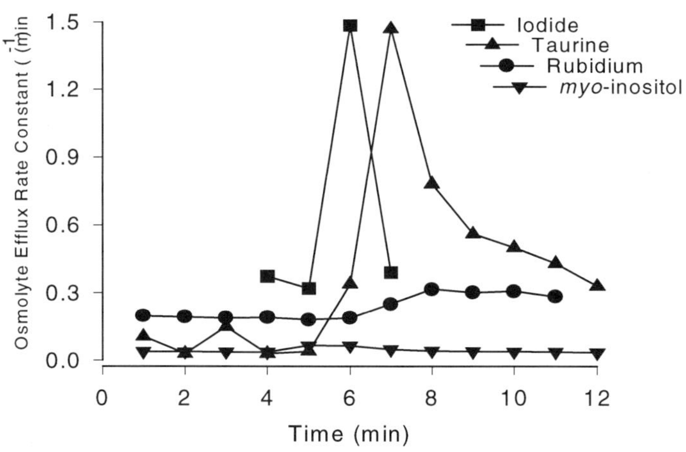

Fig. 1. Time course of osmotic swelling evoked efflux of osmolytes from cultured cerebellar astrocytes. Cells were loaded with labelled tracers, ^{86}Rb, ^{125}I, $^{3}[H]$-taurine and $^{3}[H]$-myo-inositol and after washing, perfused at 1 ml/min with isosmotic medium. At min 5th in the plot, perfusion continued with 50% hyposmotic medium.

The mechanisms leading to activation and inactivation of osmolyte fluxes are important factors modulating the efficiency of RVD but so far they are poorly understood (10). It is possible that the different pathways sense different activation signals or that activation of one of these pathways serves in turn, as signal for activating others. For instance, the sharp increase in Cl^- efflux observed immediately after swelling, which importantly contributes to the depolarization that follows osmotic swelling in cultured astrocytes (11), may in turn, activate K^+ release. This change in membrane potential may also be responsible for inactivating the Cl^- refraining the net Cl^- efflux, but it may not be a

valid explanation for amino acid pathways, particularly for those of neutral amino acids such as taurine, which at the physiological pH have no net charge and will not be that sensitive to changes in membrane potential. Also, the initial signal for activating the Cl^- efflux pathway has to be determined. The factors most commonly suggested to serve this role
and modify in accordance the RVD rate include a change in cytosolic Ca^{2+} levels, a disorganization of the actin cytoskeleton, the decrease in ion strength and the dilution of cytosolic macromolecules. We have examined some of these factors in cultured astrocytes.

The influence of Ca^{2+} has been difficult to assess. In cultured cerebellar and cortical astrocytes, hyposmotic swelling induce a raise in cytosolic Ca to which contribute both Ca^{2+} influx and release from internal stores (12,13).

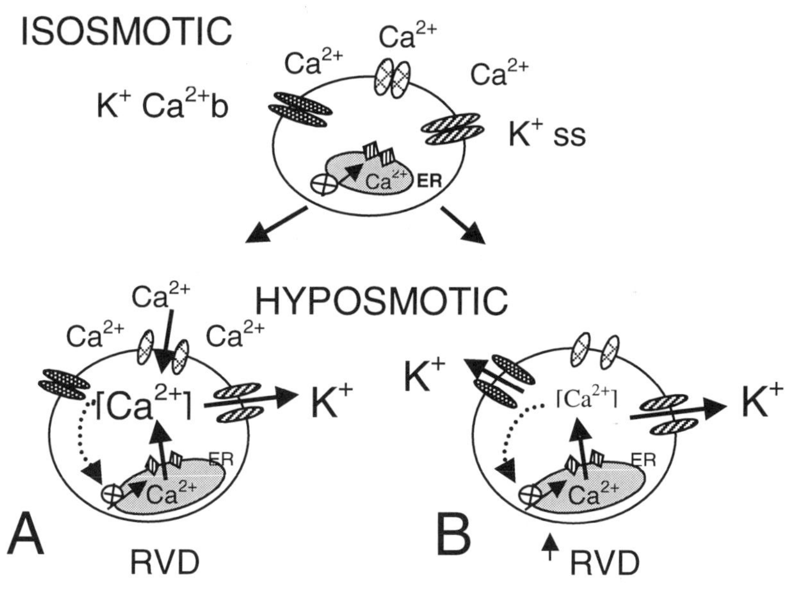

Fig. 2. Schematic model for Ca^{2+} and K^+ movements under hyposmotic conditions. Top: isosmotic condition, no movements Ca^{2+} or K^+ occur. Bottom: hyposmotic condition in the presence (A) or absence (B) of external Ca^{2+}. In (A), Ca^{2+} influx occur possibly through the capacitative pathway to replenish ER content depleted by to swelling. In (B), K^+ efflux increase by activation of the Ca^{2+} blockable K^+ pathway and RVD is accelerated. K^+ ss: swelling-sensitive K^+ pathway; $K^+Ca^{2+}b$: Ca^{2+} blockable K^+ pathway; crossed circle: Ca^{2+}-ATPase; ER: Endoplasmic Reticulum.

Despite the substantial contribution of external Ca^{2+}, RVD is not decreased but rather increased in the absence of Ca^{2+} (14). This may be due to the activation of a flow of K^+ which is found normally blocked by external Ca^{2+} as has been described in epithelial cells (15).

The superimposed release of this Ca-blockable K^+ efflux may mask any dependence of external Ca^{2+} of other osmolytes, including K^+ itself. We are currently characterizing the pharmacological properties of this Ca-blockable K^+ efflux with the aim of separate its contribution from that of the swelling activated movement. The influence of Ca^{2+} released from internal stores on RVD is also uncertain. It seems clear, from our results and those previously reported by Fischer et al. (13) that in cultured astrocytes the hyposmotic stimulus elicits a release of Ca^{2+} from the ER. The Ca^{2+} pool sensitive to osmotic swelling is also mobilized by tapsigargine (Tg), ionomycin, ATP and bradykinin, so that depletion of the pool by one of these compounds leaves no Ca^{2+} available for the osmotic response. However, RVD is unaffected by ionomycin and Tg and is accelerated by ATP, suggesting that $[Ca^{2+}]_i$ released by osmotic swelling is only an epiphenomenon not acting as a signal transduction of the change in cell volume. The Ca^{2+} release from the ER and possibly also from other endogenous sources may just be a consequence of organelle swelling which liberates Ca^{2+} by mechanical opening of Ca^{2+} channels, or by voltage-dependent channels sensing a voltage change between ER and cytosol occurring associated to swelling (16,17)(Fig.2).

The possible role of the actin cytoskeleton as sensor or signal for RVD seems unlikely, at least in cultured astrocytes, since disrupting this structure with cytochalasins B or D does not affect volume regulation nor the osmolyte fluxes (18). In astrocytes, as in many other cells, a neat change in the actin cytoskeleton organization occurs in association with changes in cell volume but we rather favor the view that these changes are necessary to permit the increase and subsequent decrease of cell volume.

There is emerging evidence suggesting that the change in cell volume itself is the signal activating and possibly inactivating as well, the efflux of Cl^- elicited by osmotic swelling (10). We have data indicating that this may be also the case for taurine. In cultured astrocytes, activation and inactivation of taurine efflux subsequent to osmotic swelling closely follows the time course of the change in cell volume. A linear correlation is observed when plotting the cell volume change and taurine efflux in different experimental situations. This correlation was found when cell volume increases by hyposmotic solutions and when is due to accumulation of K^+ and Cl^- in cells exposed to isosmotic media with KCl replacing NaCl, despite the differences in the time course under the two conditions. In astrocytes exposed to isosmotic KCl cell volume increases gradually and continuously during about one hour, to attain values about 50-60% over the initial cell volume. Under these conditions, no reduction in ion strength nor a decrease in electrolytes is associated with cell swelling and nevertheless, taurine efflux is activated. In close correlation with the increase in cell volume, a gradual and continuous efflux of taurine is observed, and the amount released during the time of exposure to KCl is similar to that released when cell volume has increased to about the same extent by hyposmotic stimulus. Therefore, the amount and rate of taurine efflux seems to follow the changes in cell volume whatever their extent and time course may be.

Taurine efflux may also be sensitive to small changes in cell volume which are barely detectable as those occurring when cells are exposed to small decreases in osmolarity. This is the process known as isovolumetric regulation. Taurine release is sensitive to these changes and is continuously released, so that after 1 h of this treatment the cell taurine pool has been reduced by 80%. This means that the correlation between taurine efflux and cell volume increase is maintained even when changes in cell volume are barely noticed. This observation is interesting since increases in cell volume observed when cells are suddenly exposed to anisosmotic are uncharacteristic of those occurring in physiological conditions. The fact that taurine but not glutamate was released during the isovolumetric regulation suggest a key role for taurine as the perfect osmolyte for the physiological control of cell volume.

References

1. Kimelberg H K, Frangakis M. (1986). Volume regulation in primary astrocyte cultures. *Adv. Biosci.* 67:177-186.

2. Pasantes-Morales H, Maar T, Morán J. (1993). Cell volume regulation in cultured cerebellar granule neurons. *J. Neurosci. Res.* 34:219-224.

3. Pasantes-Morales H, Murray R A., Lilja L, Morán J. (1994a). Regulatory volume decrease in cultured astrocytes I. Potassium- and chloride-activated permeability. *Am. J. Physiol. 266:*C165-C171.

4. Pasantes-Morales H. Murray RA, Sánchez-Olea R, Morán J. (1994b). Regulatory volume decrease in cultured astrocytes: II. Activated permeability to amino acids and polyalcohols. *Am. J. Physiol.* 266: C172-C178.

5. Pasantes-Morales H, Chacón E, Murray RA, Morán J. Properties of osmolyte fluxes activated during regulatory volume decrease in cultured cerebellar granule neurons. *J. Neurosci. Res.* 37, 720-727, 1994.

6. Pasantes-Morales H, Alavéz S, Sánchez-Olea R, Morán J. Contribution of organic

osmolytes to volume regulation in rat brain cells in culture. *Neurochem. Res.* 18, 445-452, 1993.

7. Kimelberg H K, Goderie S K, Higman S, Pang S, Waniewsky R A. (1990). Swelling-induced release of glutamate, aspartate, and taurine from astrocyte cultures. *J. Neuroscience* 10:1583-1591.

8. Strange K, Jackson PS. Selling activated organic osmolyte efflux: a new role for anion channels. *Kidney Int.* 48:994-1003, 1995.

9. Pasantes-Morales H. Cell volume regulation in brain cells: Cellular and olecular mechanisms. *Brain Met. Disease.* 11, 187-204, 1996.

10. Okada Y. Volume expansion-sensing outward-rectifying channel: fresh start to the molecular identity and volume sensor. *Am.J.Physiol.* 273: C755-C789,1997.

11. Kimelberg HK, Anderson E. Kettenmann H. (1990). Swelling-induced changes in electrophysiological properties of cultured astrocytes and oligodendrocytes. II. Whole cell currents. *Brain Res.* 529:262-268.

12. Morales-Mulia SM, Vaca L, Hernández A, Pasantes-Morales H. Cytosolic calcium changes and volume regulatory decrease in cultured cerebellar astrocytes. In preparation.

13. Fischer R, Schliess F, Haussinger D. Characterization of the hypo-osmolarity-induced Ca response in cultured rat astrocytes. *GLIA,* 20:51-58, 1997.

14. Van Driesscher W, Erlij D. Cell swelling activates a poorly selective monovalent cation channel in the apical membrane of toad urinary bladder. *Eur. J. Pharmacol.*, 428:1-8, 1994.

15. Sánchez-Olea R, Morales-Mulia M, Morán J, Pasantes-Morales H. Inhibition by dihydropyridines of regulatory volume decrease and osmolyte fluxes in cultured astrocytes is unrelated to extracellular calcium. *Neurosci. Lett.* 20, 165-168, 1995.

16. Missiaen L, De Smedt H, Parys JB, Sienaert I, Vanlingen S, Droogmans G, Nilius B, Casteels R. Hypotonically induced calcium release from intracellular stores. *J. Biol. Chem.*, 271:4601-4604, 1996.

17. Madhumita J, Minore JE, O'Neill WC. A volume-sensitive IP3 insensitive Ca store in vascular endothelial cells. *Am.J. Physiol.* 273:C316-C322, 1997.

18. Morán J, Meza I, Sabanero M, Pasantes-Morales H. Changes of the actin cytoskeleton during swelling and volume regulatory decrease in cultured astrocytes. *Am. J. Physiol.* 270, C1901-C1907, 1996.

Cell Volume Regulation: The Molecular Mechanism and Volume Sensing Machinery
Y. Okada, editor.

CELL VOLUME IN THE REGULATION OF METABOLISM, CELL PROLIFERATION AND APOPTOTIC CELL DEATH

F. Lang[a], I. Szabo[a], A. Lepple-Wienhues[a], M. Ritter[b], S. Waldegger[a], E. Gulbins[a]

[a]Department of Physiology, University of Tübingen, Gmelinstr. 5, D-72076 Tübingen, Germany

[b]Hospital for Internal Medicine, University of Innsbruck, Anichstr. 35, A-6010 Innsbruck, Austria

SUMMARY

Alterations of cell volume trigger a large number of cell volume regulatory mechanisms including ion transport, osmolyte accumulation, metabolism and expression of appropriate genes. A wealth of cellular signaling pathways link cell volume to the respective regulatory mechanisms. In the past few years it has become apparent that alterations of cell volume and the subsequent activation of cell volume regulatory mechanisms participate as part of the signaling machinery in the regulation of a variety of cellular functions including metabolism, cell proliferation and apoptotic cell death. Cell swelling inhibits degradation and favours formation of macromolecules such as proteins and glycogen, effects reversed by cell shrinkage. It is discussed that insulin exerts its antiproteolytic effect simply by inducing cell swelling due to KCl uptake through activation of $Na^+,K^+,2Cl^-$ cotransport, Na^+/H^+ exchange and Na^+/K^+ ATPase. Glucagon triggers proteolysis by cell shrinkage due to activation of ion channels. Moreover, the case will be made that stimulation of cell proliferation by expression of Ras is paralleled by increase of cell volume due to activation of $Na^+,K^+,2Cl^-$ cotransport and Na^+/H^+ exchange and that CD95 induced cell death is paralleled by activation of cell volume regulatory Cl^- channels. However, considerable additional experimentation is required to fully elucidate the role of cell volume in the control of metabolism, cell proliferation and cell death.

INTRODUCTION

As reviewed previously [1,2,3,4,5,6], a wide variety of cellular functions are sensitive to alterations of the cellular hydration state or cell volume. Alterations of cell volume modify ion and osmolyte transport across the cell membrane, a large diversity of metabolic pathways including degradation and formation of macromolecules. The effects of cell volume changes are accomplished by activation of respective transport proteins, enzymes and signaling molecules as well as altered gene expression. In recent years it became increasingly clear that the cell volume regulatory mechanisms participate in a wide variety of cellular functions, [for review see: 5]. The present brief review illustrates the participation of cell volume changes and cell volume sensitive mechanisms in the regulation of metabolism, cell proliferation and cell death. Special emphasis will be layed upon experimental evidence from the authors laboratory. For a more complete account of pertinent knowledge, the reader is referred to more extensive reviews [5,7,8].

CELL VOLUME IN THE REGULATION OF METABOLISM

Cell swelling stimulates the formation of proteins and of glycogen from amino acids and glucose phosphate, respectively, and inhibits degradation of the macromolecules. Conversely, cell shrinkage stimulates protein and glycogen degradation and inhibits protein and glycogen synthesis [9,10,11,12,13,14,15,16,17,18]. Since amino acids and glucose phosphate are osmotically more active than the respective macromolecules, these effects counteract the respective alterations of cell volume and can thus be considered cell volume regulatory mechanisms.

The effect of cell volume on protein and glycogen metabolism is exploited by hormones such as insulin and glucagon. Insulin swells hepatocytes by activation of Na^+,K^+,$2Cl^-$ cotransport and Na^+/H^+ exchange. The Na^+ thus taken up is replaced by K^+ through the action of Na^+/K^+ ATPase. Glucagon shrinks cells by activation of ion channels. The antiproteolytic effect of insulin and the proteolytic effect of glucagon in the liver are exclusively the result of the respective swelling or shrinking effect of the hormones since it is completely abolished by inhibition of the respective ion transport systems and fully mimiked by osmotic alterations of cell volume [19,20,21].

The effect of insulin and glucagon on protein synthesis and glycogen metabolism are only in part due to alterations of cell volume, the inhibitory effects of cAMP and ADH on protein synthesis, however, are similarly accounted for by their shrinking effect on hepatocytes [22].

Similar to insulin TGF-ß exerts its effect on proteolysis [23] and protein synthesis [unpublished observations in our laboratory] at least in part by inducing cell swelling. Moreover, several amino acids, such as glycin and glutamine modify proteolysis by altering cell volume [15,24].

The antiproteolysis may partially be due to the alkalinizing effect of cell swelling on acidic cellular compartments [25,26,27,28,29,30,31,32], since the proteolytic enzymes in those compartments are highly sensitive to ambient pH [33].

Besides its influence on macromolecule metabolism, cell swelling inhibits glycolysis and gluconeogenesis and stimulates flux through the pentose phosphate pathway, NADPH production [8,34,35] and glutathione (GSH) formation and efflux into blood [36]. Furthermore, cell swelling stimulates glycine oxidation [37], glutamine breakdown [38], formation of NH_4^+ and urea from amino acids [39], ketoisocaproate oxidation [37] and lipogenesis from glucose [40].

Cell swelling stimulates and cell shrinkage inhibits flux through the pentose phosphate pathway and formation of NADPH [34], a substrate for glutathione reductase. Accordingly, cell swelling favours production of GSH and thus provides protection against peroxides [34]. On the other hand, cell swelling presumably favours the formation of superoxide by NADPH oxidase which is in turn inhibited by osmotic cell shrinkage [41,42,43,44,45,46]. Furthermore, components of the cytosolic burst oxidase have been found to be dissociated by high osmolarity [47].

Cell volume influences the expression of a variety of genes. Cell shrinkage stimulates expression of enzymes or transporters accomplishing cellular formation or accumulation of osmolytes, such as the aldose reductase, and the Na^+ coupled transporters for betaine, taurine, inositol and amino acids as well as Na^+,K^+,$2Cl^-$ cotransport [for review see: 1,2,5,6]. Cell shrinkage stimulates the expression of the cell volume regulated kinase h-sgk possibly triggering some mechanism(s) of regulatory cell volume increase [48]. Other genes stimulated by cell shrinkage comprise ClC-K1, P-glycoprotein, heat shock proteins, the immediate early genes Egr-1 and c-fos, cycloxygenase-2, phosphoenolpyruvate carboxykinase (PEPCK), tyrosine aminotransferase, aB crystallin, laminin B_2 and matrix proteins. Genes stimulated by cell swelling include ß-actin, tubulin, immediate early gene c-jun, ornithine decarboxylase, arginine succinate lyase and tissue plasminogen activator [1,5].

CELL VOLUME AND CELL PROLIFERATION

A wide variety of mitogenic factors activate the Na^+/H^+ exchanger and many factors stimulate $Na^+,K^+,2Cl^-$ cotransport [for review see: 7,49,50,51] leading to increase of cell volume [for review see: 5]. The increase of cell volume coincides with the entry of fibroblasts from the G_1 into the S phase [52], which is accompanied by inhibition of K^+ channels [53]. Accordingly, hypertonic shrinkage inhibits [54,55,56,57,58,59] and slight osmotic cell swelling accelerates [60] cell proliferation. If exposed to excessive ionic strength cells have to overcome cell shrinkage by cellular accumulation of osmolytes in order to proliferate [59,61].

The expression of *ras* oncogene in fibroblasts is similarly paralleled by an increase of cell volume [62,63,64]. The increase of cell volume requires oscillations of intracellular Ca^{++} activity [65], triggered by bradykinin, bombesin or serum in those cells [62,63,66,67,68,69]. Initially, the Ca^{++} oscillations cause rapid transient cell shrinkage due to activation of Ca^{++} sensitive K^+ channels [70,71]. This initial cell shrinkage is, however, followed by a sustained increase of cell volume secondary to a depolymerization of the actin filaments and subsequent activation of Na^+/H^+ exchange and $Na^+,K^+,2Cl^-$ cotransport [65,72,73,74]. Cell shrinkage, on the other hand, favours the occurrence of the Ca^{++} oscillations [75].

Activation of both, Ca^{++} oscillations and K^+ channels [66,71,76,77,78,79,80,81,82, 83,84,85] as well as Na^+/H^+ exchange and $Na^+,K^+,2Cl^-$ cotransport [63,64] are required for stimulation of cell proliferation by *ras* oncogene and diverse mitogenic stimuli in a wide variety of cells.

The mechanisms linking cell volume changes to the machinery of cell proliferation are poorly understood. Cell swelling stimulates the protein kinases ERK-1 and ERK-2 [86], which are involved in the regulation of cell cycle [87].

CELL VOLUME AND APOPTOSIS

One of the hallmarks of apoptotic cell death is cell shrinkage [88,89] which may in some cells be secondary to an increase of intracellular Ca^{++} activity [90,91] and a subsequent activation of Ca^{++} sensitive ion channels [92,93].

Massive cell shrinkage, by doubling of extracellular osmolarity has in turn been shown to trigger apoptosis [94,95]. Furthermore, the resistance of cells to apoptosis following an osmotic shock has been correlated with their ability to cell volume regulation [94].

Triggering of the Fas (CD95) receptor induces apoptosis in Jurkat lymphocytes [96,97], an effect paralleled by activation of the outwardly rectifying Cl^- channel (ORCC) [98]. The same channel is activated by cell swelling and is required for regulatory cell volume decrease [99]. Activation of the channel by cell swelling and by the Fas receptor is mediated by the tyrosine kinase lck^{56} [98,99]. Activation of the channel were expected to induce cell shrinkage. However, cell volume does not decrease within the first 60 minutes following Fas triggering even though the

Cl^- channel is activated within few minutes [98]. Instead, the activation of the Cl^- channel leads to marked cytosolic acidification [98] which may be important for induction of apoptosis [100].

The lack of early cell shrinkage may be due to inhibition of the n-type K^+ channel or Kv1.3, an effect again mediated by lck^{56} [101,102]. The Kv1.3 is the cell volume regulatory K^+ channel in lymphocytes [103]. The inhibition of early cell shrinkage may be important for induction of apoptosis, since moderate osmotic cell shrinkage has been observed to inhibit CD95 induced cell death of Jurkat lymphocytes [104].

With delay of more than one hour, triggering of the Fas receptor stimulates the release of taurine [105]. At this point the loss of osmolytes favours cell shrinkage.

How cell volume initiates or inhibits apoptotic cell death, however, is largely elusive.

Massive cell shrinkage has been shown to elicit clustering and ligand-independent activation of the TNF-receptor triggering the signaling cascade for TNF induced apoptotic cell death [106]. Moreover, cell shrinkage stimulates the mitogen activated protein kinase (MAPK) pathway leading to activation of JNK via the MAPK kinase (MKK4) [106,107,108,109]. Since JNK and p38 kinase, another target of MKK4, have been invoked in the triggering of apoptosis [110], triggering of this pathway could indeed induce apoptosis. On the other hand MKK4 has been postulated to protect from apoptosis [111].

The mechanims mediating the protective effect of moderate cell shrinkage against Fas-induced cell death is similarly illdefined. Apparently, cell shrinkage interferes at some point with the signalling cascade of the Fas-receptor [104], which involves activation of caspases with subsequent stimulation of sphingomyelinase, ceramide formation, activation of Ras, PI3 kinase, Rac, MKK4 and JNK or p38 kinase [96,97,110,112,113]. The activation of Rac triggers formation of superoxides, presumably by the NADPH oxidase [104]. Cell shrinkage apparently does not interfere with the cascade leading to Ras or Rac activation but prevents the formation of superoxides, since the decrease of GSH is blunted in shrunken lymphocytes [104]. As outlined above (see 2), cell shrinkage inhibits glucose flux through the pentose phosphate pathway decreasing the availability of NADPH and inhibits superoxide formation through NADPH oxidase.

ACKNOWLEDGEMENTS

Research in the authors laboratories has been supported by the DFG (La 315/4-3), the EU (contract number ERBCHRXCT 94-0595) and the Baxter Foundation.

REFERENCES

1 Burg MB, Kwon ED, Kultz D. FASEB J 1996; 10(14): 1598-1606.
2 Garcia-Perez A, Burg MB. Physiol Rev 1991; 71: 1081-1115.
3 Hoffmann EK, Dunham PB. Int Rev Cytology 1995; 161: 173-262.
4 Kinne RKH, Czekay R-P, Grunewald JM, Mooren FC, et al. Renal Physiol Biochem 1993; 16: 66-78.
5 Lang F, Busch GL, Ritter M, Völkl H, et al. Physiol Rev 1998a (in press).
6 Lang F, Busch GL, Völkl H. Cell Physiol Biochem 1998b; 8:1-45.
7 Ritter M, Wöll E. Cell Physiol Biochem 1996; 6(5): 245-270.

8 Häussinger D, Newsome W, Vom Dahl S, Stoll B, et al. Biochem Soc Trans 1994; 22: 497-502.
9 Baquet A, Hue L, Meijer AJ, Van Woerkom GM, et al. J Biol Chem 1990; 265: 955-959.
10 Häussinger D, Lang F. Cell Physiol Biochem 1991a; 1: 121-130.
11 Häussinger D, Lang F. Biochim Biophys Acta 1991b; 1071: 331-350.
12 Häussinger D, Hallbrucker C, Vom Dahl S, Lang F, et al. Biochem J 1990a; 272: 239-242.
13 Häussinger D, Hallbrucker C, Vom Dahl S, Decker S, et al. FEBS Lett 1991; 283: 70-72.
14 Häussinger D, Roth E, Lang F, Gerok W. Lancet 1993; 341: 1330-1332.
15 Hallbrucker C, Vom Dahl S, Lang F, Häussinger D. Eur J Biochem 1991a; 197: 717-724.
16 Lang F, Stehle T, Häussinger D. Pflügers Arch 1989; 413: 209-216.
17 Meijer AJ, Baquet A, Gustafson L, Van Woerkom GM, et al. J Biol Chem 1992; 267: 5823-5828.
18 Peak M, Al-Habouri M, Agius L. Biochem J 1992; 282: 797-805.
19 Vom Dahl S, Hallbrucker C, Lang F, Gerok W, et al. Biochem J 1991a; 278: 771-777.
20 Vom Dahl S, Hallbrucker C, Lang F, Häussinger D. Biochem J 1991b; 280: 105-109.
21 Hallbrucker C, Vom Dahl S, Lang F, Gerok W, et al. Eur J Biochem 1991b; 199: 467-474.
22 Stoll B, Gerok W, Lang F, Häussinger D. Biochem J 1992; 287: 217-222.
23 Ling H, Vamvakas S, Busch GL, Kulzer P, et al. Am J Physiol 1995; 269:F911-F917.
24 Vom Dahl S, Häussinger D. J Nutr 1996; 126: 395-402.
25 Busch GL, Schreiber R, Dartsch PC, Völkl H, et al. Proc Natl Acad Sci USA 1994; 91: 9165-9169.
26 Busch GL, Lang H-J, Lang F. Pflügers Arch 1996a; 431: 690-696.
27 Busch GL, Günther E, Hewig B, Zrenner E, et al. Cell Physiol Biochem 1996b; 6: 185-194.
28 Busch GL, Wiesinger H, Gulbins E, Wagner HJ, et al. Biochim Biophys Acta 1996c; 1285: 212-218.
29 Busch GL, Völkl H, Haller M, Siemen D, et al. Cell Physiol Biochem 1997; 7: 25-34.
30 Lang F, Busch GL, Völkl H, Häussinger D. Biochem Soc Trans 1994; 22: 502-504.
31 Völkl H, Rehwald W, Waitz W, Häussinger D, et al. Cell Physiol Biochem 1993a; 3: 28-33.
32 Völkl H, Friedrich F, Häussinger D, Lang F. Biochem J 1993b; 295: 11-14.
33 Mortimore GE, Pösö AR. Ann Rev Nutr 1987; 7: 539-564.
34 Saha N, Stoll B, Lang F, Häussinger D. Eur J Biochem 1992; 209: 437-444.
35 Saha N, Schreiber R, Vom Dahl S, Lang F, et al. Biochem J 1993; 296: 701-707.
36 Häussinger D, Lang F, Bauers K, Gerok W. Eur J Biochem 1990c; 193: 891-898.

37 Häussinger D, Stoll B, Morimoto Y, Lang F, et al. Biol Chem Hoppe-Seyler 1992; 373: 723-734.
38 Häussinger D, Lang F, Bauers K, Gerok W. Eur J Biochem 1990b; 188: 689-695.
39 Häussinger D, Lang F. J Cell Biochem 1990; 43: 355-361.
40 Baquet A, Lavoinne A, Hue L. Biochem J 1991; 273: 57-62.
41 Kataoka S, Fujita Y. Nippon Hinyokika Gakkai Zasshi 1991; 82: 16-23.
42 Kuchkina NV, Orlov SN, Pokudin NI, Chuchalin AG. Biull Eksp Biol Med 1993a; 115: 360-362.
43 Kuchkina NV, Orlov SN, Pokudin NI, Chuchalin AG. Experientia 1993b; 49: 995-997.
44 Manara FS, Chin J, Schneider DL. J Leukoc Biol 1991; 49: 489-498.
45 Matsumoto T, Van der Auwera P, Watanabe Y, Tanaka M, et al. Urol Res 1991; 19: 223-227.
46 Takahashi K, Matsumoto T, Kubo S, Haraoka M, et al. J Urol 1994; 152: 1622-1625.
47 Iyer SS, Pearson DW, Nauseef WM, Clark RA. J Biol Chem 1994; 269: 22405-22411.
48 Waldegger S, Barth P, Raber G, Lang F. Proc Natl Acad Sci USA 1997; 94: 4440-4445.
49 Bianchini L, Grinstein S. In: Lang F, Häussinger D, eds. Advances in Comparative and Environmental Physiology. Berlin, Heidelberg: Springer-Verlag, Vol. 14, 1993; 249-270.
50 Grinstein S, Rotin D, Mason JM. Biochim Biophys Acta 1989; 988: 73-97.
51 Palfrey HC, O'Donnell ME. Cell Physiol Biochem 1992; 2: 293-307.
52 Pendergrass WR, Angello JC, Kirschner MD, Norwood TH. Exp Cell Res 1991; 192: 418-425.
53 Day ML, Pickering SJ, Johnson MH, Cook DI. Nature 1993; 365: 560-562.
54 Albertyn J, Hohmann S, Thevelein JM, Prior BA. Mol Cell Biol 1994; 14: 4135-4144.
55 Higgins CF, Cairney J, Stirling DA, Sutherland L, et al. Trends Biochem Sci 1987; 12: 339-344.
56 Miyoshi K, Funahashi H, Okuda K, Niwa K. J Reprod Fertil 1994; 100: 21-26.
57 Petronini PG, De-Angelis EM, Borghetti P, Borghetti AF, et al. Biochem J 1992; 282: 69-73.
58 Robbins E, Pederson T, Klein P. J Cell Biol 1970; 44: 400-416.
59 Yancey PH, Burg MB. Am J Physiol 1990; 258: R198-R204.
60 Anbari K, Schultz RM. Mol Reprod Dev 1993; 35: 24-28.
61 Lawitts JA, Biggers JD. Mol Reprod Dev 1992; 31: 189-194.
62 Lang F, Ritter M, Wöll E, Weiss H, et al. Pflügers Arch 1992a; 420. 424-427.
63 Lang F, Ritter M, Wöll E, Bichler I, et al. Cell Physiol Biochem 1992b; 2: 213-224.
64 Meyer M, Maly K, Überall F, Hoflacher J, et al. J Biol Chem 1991; 266: 8230-8235.
65 Dartsch PC, Ritter M, Gschwentner M, Lang H-J, et al. Eur J Cell Biol 1995; 67: 372-378.
66 Lang F, Friedrich F, Kahn E, Wöll E, et al. J Biol Chem 1991; 266: 4938-4942.

67 Waldegger S, Pinggera G, Kloiber K, Ritter M, et al. Cell Physiol Biochem 1993; 3: 89-96.
68 Wöll E, Weiss H, Waldegger S, Lang F. Eur J Pharmacol 1992a; 212: 105-107.
69 Wöll E, Waldegger S, Lang F, Maly K, et al. Pflügers Arch 1992b; 420: 208-212.
70 Ritter M, Wöll E, Häussinger D, Lang F. FEBS Lett 1992; 307: 367-370.
71 Wöll E, Ritter M, Scholz W, Häussinger D, et al. FEBS Lett 1993; 322: 261-265.
72 Dartsch PC, Ritter M, Häussinger D, Lang F. Eur J Cell Biol 1994; 63: 316-325.
73 Ritter M, Wöll E, Haller T, Dartsch PC, et al. Eur J Cell Biol 1997a; 72: 222-228.
74 Ritter M, Dartsch PC, Waldegger S, Haller T, et al. Biochim Biophys Acta 1997b; 1358: 20-30.
75 Ritter M, Wöll E, Waldegger S, Häussinger D, et al. Pflügers Arch 1993; 423: 221-224.
76 Amigorena S, Choquet D, Teillaud JL, Korn H, et al. J Immunol 1990; 144: 2038-2045.
77 Leonard RJ, Garcia ML, Slaughter RS, Reuben JP. Proc Natl Acad Sci USA 1992; 89: 10094-10098.
78 Lewis RS, Cahalan MD. Trends Neurosci 1988; 11: 214-218.
79 Nilius B, Wohlrab W. J Physiol Lond 1992; 445: 537-548.
80 Puro DG, Roberge F, Chan CC. Invest Ophthalmol Vis Sci 1989; 30: 521-529.
81 Schuller HM, Orloff M, Reznik GK. Carcinogenesis 1991; 12: 2301-2303.
82 Shultz PJ, Raij L. Hypertens 1990; 15: 176-180.
83 Teulon J, Ronco PM, Geniteau-Legendre M, Baudouin B, et al. J Cell Physiol 1992; 151: 113-125.
84 Vandewalle A, Vuillemin T, Teulon J, Boudouin B, et al. J Cell Physiol 1993; 154: 466-477.
85 Xu B, Wilson BA, Lu L. Am J Physiol 1996; 40: C2037-C2044.
86 Schliess F, Schreiber R, Häussinger D. Biochem J 1995; 309: 13-17.
87 Boulton TG, Yancopoulos GD, Gregory JS, Slaughter C, et al. Science 1990; 249: 64-67.
88 Wyllie AH, Kerr JF, Currie AR. Int Rev Cytol 1980; 68: 251-306.
89 Kerr JF, Wyllie AH, Currie AR. Br J Cancer 1972; 26: 239-257.
90 Oshimi Y, Miyazaki S. J Immunol 1995; 154: 599-609.
91 Trump BF, Berezesky IK. FASEB J 1995; 9: 219-228.
92 Barbiero G, Duranti F, Bonelli G, Amenta JS, et al. Exp Cell Res 1995; 217: 410-418.
93 Beauvais F, Michel L, Dubertret L. J Leukoc Biol 1995; 57: 851-855.
94 Bortner CD, Cidlowski JA. Am J Physiol 1996; 40: C950-C961.
95 Matthews CC, Feldman EL. J Cell Physiol 1996; 166: 323-331.
96 Gulbins E, Bissonnette R, Mahboubi A, Martin S, et al. Immunity 1995; 2: 341-351.
97 Gulbins E, Coggeshall KM, Brenner B, Schlottmann K, et al. J Biol Chem 1996a; 271: 26389-26394.
98 Szabó I, Kaba NK, Zoratti M, Lepple-Wienhues A, et al. Proc Natl Acad Sci USA 1998 (submitted)
99 Lepple-Wienhues A, Szabò I, Laun T, Kaba NK, et al. Pflügers Arch 1998 (in press).

100 Gottlieb RA, Nordberg J, Skowronski E, Babior BM. Proc Natl Acad Sci USA 1996; 93:654-658.
101 Gulbins E, Szabò I, Baltzer K, Lang F. Proc Natl Acad Sci USA 1997a; 94: 7661-7666.
102 Szabò I, Gulbins E, Apfel H, Zhang X, et al. J Biol Chem 1996; 271: 20465-20469.
103 Deutsch C, Chen LQ. Proc Natl Acad Sci USA 1993; 90: 10036-10040.
104 Gulbins E, Welsch J, Lepple-Wienhues A, Heinle H, et al. Biochem Biophys Res Commun 1997b; 236: 517-521.
105 Lang F, Madlung J, Gulbins E. Pflügers Arch 1998 (in press).
106 Rosette C, Karin M. Science 1996; 274: 1194-1197.
107 Matsuda S, Kawasaki H, Moriguchi T, Gotoh Y, et al. J Biol Chem 1995; 270: 12781-12786.
108 Moriguchi T, Kawasaki H, Matsuda S, Gotoh Y, et al. J Biol Chem 1995; 270:12969-12972.
109 Tokiwa G, Dikic I, Lev S, Schlesinger J. Science 1996; 273: 792-794.
110 Xia Z, Dickens M, Raingeaud J, Davis RJ, et al. Science 1995; 270: 1326-1331.
111 Nishina H, Fischer KD, Radvanyi L, Shahinian A, et al. Nature 1997; 385: 350-353.
112 Gulbins E, Brenner B, Schlottmann K, Welsch J, et al. Immunology 1996b; 89: 205-212.
113 Gulbins E, Szabò I, Lang F. Cell Physiol Biochem 1996c; 6: 361-375.

PROTEIN TYROSINE PHOSPHORYLATION IS A PREREQUISITE FOR ACTIVATION OF THE OSMO-SENSITIVE ANION EFFLUX

Hypo-osmotic swelling of Intestine 407 cells is accompanied by a rapid and transient increase in tyrosine phosphorylation of several proteins, including the p125 focal adhesion kinase ($p125^{FAK}$) and the Erk-1/Erk-2 map kinases [3, 4]. To investigate a possible involvement of tyrosine kinases in the regulation of osmo-sensitive anion channels, hypotonicity-provoked $^{125}I^-$ efflux was quantitated from isotope-loaded cultures treated with either vanadate, a potent inhibitor of phospho-tyrosine phosphatases, or herbimycin A, a specific inhibitor of tyrosine kinases. As shown in Fig.2, vanadate treatment of the cells increased the $^{125}I^-$ efflux induced by a mild (20%) hypotonic shock by approx. 2-3 fold, but did not affect the response triggered by a saturating stimulus (30%). Notably, in vanadate-treated cultures, osmotic cell swelling was largely prevented after mild hypotonic stimulation, but not after stimulation with hypo-osmotic media that induce a maximal ionic response [3]. Pre-treatment of the cells with herbimycin A strongly reduced the osmo-shock-induced efflux of $^{125}I^-$. The inhibition was observed in control as well as in vanadate pre-treated cells and within

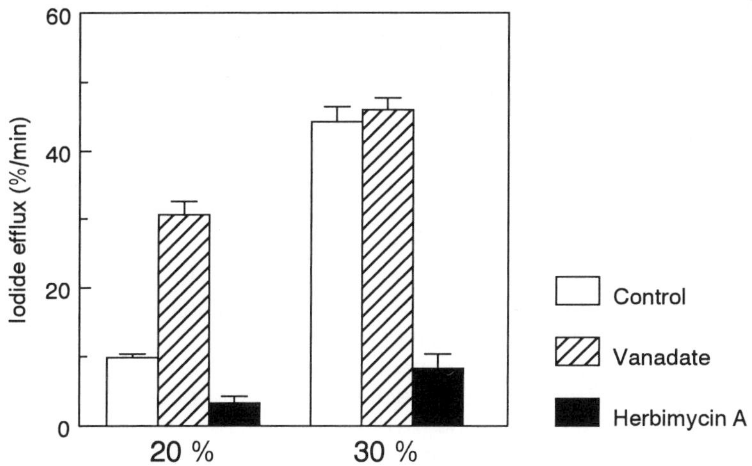

Figure 2. Sensitivity of the cell swelling-induced anion efflux to vanadate and herbimycin A.
Isotope efflux from control and vanadate- (200µM, 5 min) or herbimycin A- (1µM, 48h) treated cultures after hypo-osmotic stimulation (20% or 30% reduction in tonicity). Data are expressed as peak iodide efflux (mean ± SD for n = 3).

a broad range of hypotonicity (10 - 30%). Taken together, these results lead to the conclusion that protein tyrosine kinases phosphorylation is an essential step in, or a prerequisite for the activation of osmo-sensitive anion channels.

OSMOTIC CELL SWELLING ACTIVATES MEMBERS OF THE MAP KINASE FAMILY

Osmotic activation of map kinases has now been reported for a variety of mammalian cell types and for yeast [1,2]. For Intestine 407 cells, activation of both the Erk-1/Erk-2 map kinases, implicated in cell proliferation and differentiation, as well as the stress-induced p38 and JNK map kinases has been observed [3, 5]. Activation of Erk-1/Erk-2 and p38 is rapid and proceeds with similar kinetics as the opening of compensatory Cl^- channels. In contrast, activation of JNK is much slower and occurs only after a distinct lag phase of 10-20 minutes, arguing against a direct role of JNK in the RVD response (B.C.Tilly, unpublished results).

In yeast, both the Hog1p and Mpk1p members of the map kinase family are critically involved in cell volume regulation [6, 7]. The osmo-sensing mechanism leading to the activation of the Hog1p, which is highly homologous to p38 map kinase, has recently been elucidated [8, 9]. The osmosensor belongs to a so called "two-component signal-transduction system" and consists of a transmembrane histidine kinase Sln1p (the "sensor") and a cytosolic response regulator (Ssk1p). As a rapid activation of Erk-1/Erk-2 and p38 map kinases is among the initial signalling events after osmotic cell swelling of Intestine 407 cells, it is tempting to propose a model in which map kinases play an important role in the RVD response. Inhibition of the p38 map kinase cascade, however, using the specific p38 inhibitor SB-203580, did not block the osmo-sensitive anion efflux and, *vice versa*, activation of p38 map kinase by cytokines and anisomycin did not increase the anion efflux [5]. On the other hand, stimulation of the cells with EGF was found not only to activate Erk-1/Erk-2 but also to strongly potentiate the hypotonicity-induced ion efflux [3]. To investigate a possible role of Erk-1/Erk-2 in the regulation of osmo-sensitive Cl^- channels, dominant negative p21-Ras (Ras-N17), a mutated species of the small GTP binding protein involved in growth factor-induced Erk-1/Erk-2 activation, was introduced into the cells. Whereas, in Ras-N17 infected cells, the hypotonicity-provoked Erk-1/Erk-2 activation was completely inhibited, neither the cell swelling-induced activation of the $^{125}I^-$ efflux (Fig.3), nor its potentiation by EGF (results not shown) was reduced. Collectively, these results argue against a direct role for members of the map kinase family in the regulation of osmo-sensitive anion channels. So far, little is known about a possible other physiological role of map kinase activation during the RVD response. However, since the involvement of map kinases in transcription regulation is now well established, it is tempting to postulate a role for Erk-1/Erk-2/p38 in restoring cellular homeostasis and promoting long term survival.

REGULATION OF THE OSMO-SENSITIVE ANION EFFLUX BY P21-RHO

Hypo-osmotic stimulation of Intestine 407 cells was accompanied by a transient increase in F-actin and a reorganization of the cytoskeleton [4]. Numerous short and poorly organized fibers were observed within the first minute of osmotic stimulation, causing a prominent increase in actin stress fiber formation after 2 minutes. Subsequently, stress fiber formation declined and the amount of F-actin returned to pre-stimulatory levels after 5 - 15 minutes. The increase in stress fiber formation was accompanied by a transient loss in apical F-actin, which is part of the terminal web.

In many cell types, stimulus-induced reorganization of the actin cytoskeleton involves the activation of p21-Rho and/or p21-Rac, both members of the Ras family of small GTPases. Using recombinant *Clostridium botulinum* C3 exoenzyme to inhibit p21-Rho, a large reduction of the hypotonicity-induced anion efflux was observed (Fig.3). This inhibition was not observed when mutant enzyme (C3-E173Q) devoid of ADP-ribosyltransferase activity was used, suggesting a role for p21-Rho in the regulation of osmo-sensitive Cl^- channels, most plausibly through the induction of

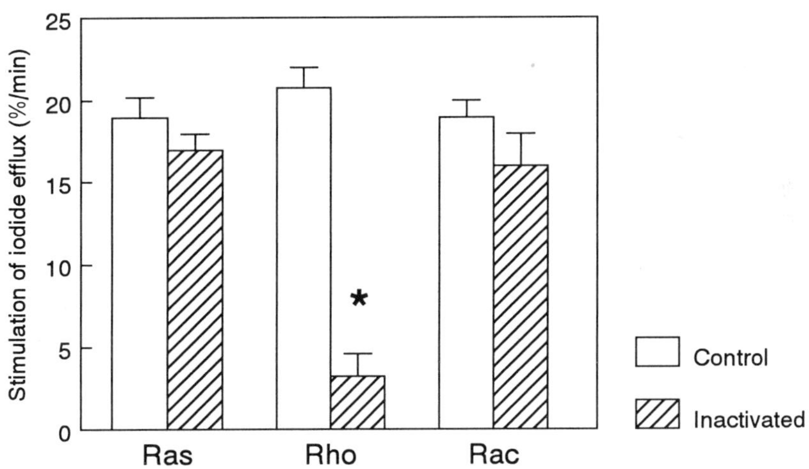

Figure 3. Role of Ras-related G-proteins in the regulation of cell swelling-activated anion efflux.

Cells were either infected with recombinant vaccinia virus expressing dominant negative Ras-N17 or Rac-N17 or treated with *C.botulinum* C3 exoenzyme (50 µg/ml, 48h) to inactivate Ras, Rac and Rho, respectively (hatched bars). Thereafter, the hypotonicity-provoked iodide efflux (30% hypotonicity) was quantitated (data are expressed as mean ± SD, n=3). Open bars represent cell swelling-induced iodide efflux in control cultures.

cytoskeletal alterations [4]. In contrast, introducing dominant negative Rac-N17 into the cells, similar to Ras-N17, did not affect the volume-sensitive efflux (Fig.3). At least two possible mechanisms may underlie the regulation of ionic channels by cytoskeletal re-arrangements. First, generation of actin filaments of a specific size may be crucial for activation, a notion supported by our observations that pre-treatment of the cells with cytochalasin B potentiated the osmo-sensitive $^{125}I^-$ efflux [4]. Alternatively, modulation of signal transduction pathways associated with cytoskeletal re-arrangements, such as the one involving $p125^{FAK}$, could also be involved.

The transient re-organization of F-actin during the RVD response, suggests a potential role for p38 map kinase. Through its downstream target, MAPKAP kinase 2, activation of p38 leads to phosphorylation of the small heat shock proteins Hsp25 and Hsp27 [5]. Both Hsp25 and Hsp27 have been implicated as molecular chaperones and as inhibitors of F-actin polymerization. In contrast to their chaperone function, which has been shown to be independent of protein phosphorylation, inhibition of actin polymerization occurs only in the presence of non-phosphorylated Hsp25 and was abolished after its phosphorylation [10].

Cell swelling-induced remodelling of the cytoskeleton also coincided with a transient, C3 exoenzyme-sensitive increase in phosphatidylinositol-3-kinase (PtdIns-3-kinase) activity [4]. Although the cell swelling-induced increase in PtdIns-3-kinase activity is modest as compared to its stimulation by EGF, pre-treatment of the cells with wortmannin, an inhibitor of PtdIns-3-kinase, largely abolished the hypotonicity-induced $^{125}I^-$ efflux [4].

CONCLUSIONS

Osmotic cell swelling leads to a rapid phosphorylation and/or activation of a number of signalling enzymes concomitant with an increase in the conductance of compensatory ionic channels (Fig.4). In contrast to the activation of the map kinases, which apparently does not play a causative role in ionic channel regulation, activation of p21-Rho and PtdIns-3-kinase was found to be a pre-requisite for the proper activation of osmo-sensitive anion channels in Intestine 407 cells. Because both enzymes play a role as regulators of the actin cytoskeleton, F-actin itself or rearrangements in its cellular distribution may be involved in the regulation of cell swelling-induced Cl^- channels.

PtdIns-3-kinase may, apart from its possible role in cell swelling induced remodelling of the cytoskeleton, serve an additional function. Through its product Ins-3,4-P_2, this lipid kinase has been implicated in the activation of protein kinase B (PKB or Akt), a serine-threonine kinase involved in the stimulation of glycogen synthase kinase-3 [11]. Recent observations suggest that PKB is not only involved in increasing glycogen synthesis, but also mediates cell survival and prevents programmed cell death [11]. This raises the interesting possibility that PtdIns-3-kinase/PKB may serve multiple functions during the RVD response: (1) activation of Cl^- channels, either directly or through the cytoskeleton; (2) reducing cellular osmolytes by promoting glycogen synthesis and (3) increasing cell survival.

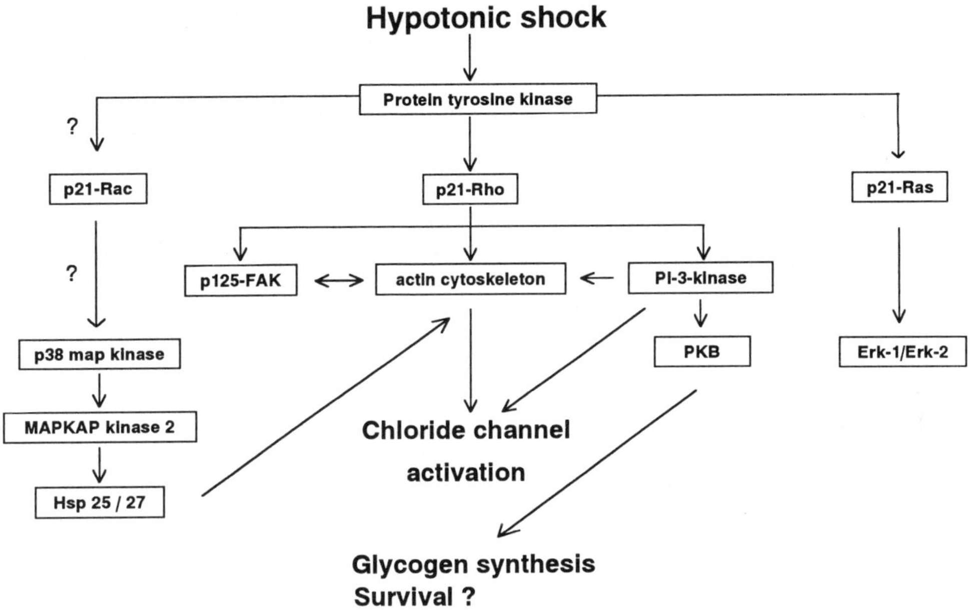

Figure 4. Cellular signalling pathways activated by osmotic cell swelling.

Importantly, hormonal activation of signalling molecules like p21-Rho, $p125^{FAK}$ and PtdIns-3-kinase under isotonic conditions, as observed in the presence of EGF and/ or Ca^{2+}-mobilizing hormones, is clearly not sufficient to activate volume-sensitive Cl^- channels [3,12], indicating that an additional factor, perhaps directly related to membrane stretch, is also essential. Linking the signal transduction pathways involved in Cl^- channel activation to an independent putative 'volume sensor' has two important physiological consequences. First, activation of the channels occurs only during cell swelling, thereby preserving the specificity of the response, and second, by coupling the volume response to pathways activated upon hormonal stimulation, small changes in cell volume which could take place during hormonal stimulation, are more adequately corrected.

REFERENCES

1. Okada Y. Volume expansion-sensing outward-rectifier Cl- channel: fresh start to the molecular identity and volume sensor. Am J Physiol 1997; 273: C755-C789.

2. Nilius B, Eggermont J, Voets T, Buyse G, Manolopoulos V, Droogmans G. Properties of volume-regulated anion channels in mammalian cells. Prog Biophys Mol Biol 1997; 68: 1-61.
3. Tilly BC, van den Berghe N, Tertoolen LGJ, Edixhoven MJ, de Jonge HR. Protein tyrosine phosphorylation is involved in osmoregulation of ionic conductances. J Biol Chem 1993; 268: 19919-19922.
4 Tilly BC, Edixhoven MJ, Tertoolen LGJ, Morii N, Saito Y, Narumiya S, de Jonge HR. Activation of the osmo-sensitive chloride conductance involves $p21^{rho}$ and is accompanied by a transient reorganization of the F-actin cytoskeleton. Mol Biol Cell 1996; 7: 1419-1427.
5. Tilly BC, Gaestel M, Engel K, Edixhoven MJ, de Jonge HR. Hypo-osmotic cell swelling activates the p38 map kinase signalling cascade. Febs Letters 1996; 395: 133-136.
6. Brewster JL, de Valoir T, Dwyer ND, Winter E, Gustin MC. An osmosensing signal transduction pathway in yeast. Science 1993; 259: 1760-1763.
7. Davenport KR, Sohaskey M, Kamada Y, Levin DE, Gustin MC. A second osmosensing signal transduction pathway in yeast. Hypotonic shock activates the PKC1 protein kinase-regulated cell integrity pathway. J Biol Chem 1995; 270: 30157-30161.
8. Maeda T, Wurgler-Murphy SM, Saito H. A two-component system that regulates an osmosensing MAP kinase cascade in yeast. Nature 1994; 369: 242-245.
9. Posas F, Wurgler-Murphy SM, Maeda T, Witten EA, Thai TC, Saito H. Yeast HOG1 MAP kinase cascade is regulated by a multistep phosphorelay mechanism in the SLN1-YPD1-SSK1 "two-component" osmosensor. Cell 1996; 86: 865-875.
10. Miron T, Van Compernolle K, Van de Kerckhove J, Wilchek M, Geiger B. A 25-kD inhibitor of actin polymerization is a low molecular mass heat shock protein. J Cell Biol 1991; 114: 255-261.
11. Franke TF, Kaplan DR, Cantley L.C. PI3K: downstream AKTion blocks apoptosis. Cell 1997; 88: 435-437, 1997.
12. Tilly BC, Edixhoven MJ, van den Berghe N, Bot AGM, de Jonge HR. Ca^{2+}-mobilizing hormones potentiate hypotonicity-induced activation of ionic conductances in Intestine 407 cells. J Am Physiol 1994; 267: C1271-C1278, 1994.

Cell Volume Regulation: The Molecular Mechanism and Volume Sensing Machinery
Y. Okada, editor.

Dual calcium response to cell volume change in aortic endothelium.

Masahiro Oike, Chiwaka Kimura, Tetsuya Koyama and Yushi Ito

Department of Pharmacology, Faculty of Medicine, Kyushu University, Fukuoka, 812-8582, Japan.

INTRODUCTION

Vascular endothelium is known to produce various biologically active substances such as nitric oxide, endothelin and growth factors in response to mechanical stress (for a review, see 1). It is generally agreed that the elevation of intracellular Ca^{2+} concentration ($[Ca^{2+}]_i$) is necessary to produce and secrete these substances. Therefore, the investigation of mechanosensitive Ca^{2+} mobilization would have significant importance for understanding endothelial physiology. The stretch-activated non-selective cation channel is an important candidate for such Ca^{2+} mobilization in endothelium (1). Another possible source for mechanosensitive $[Ca^{2+}]_i$ elevation is Ca^{2+} release from intracellular store sites. We have previously reported that mechanical stress induces Ca^{2+} release by the production of arachidonic acid in endothelial cells from human umbilical vein (6). Another possible mechanism that has also been reported is that mechanical stress induces the secretion of agonists from the endothelium such as ATP, substance P or prostaglandin which could also induce Ca^{2+} release (2). Aortic endothelium is exposed to a stronger mechanical stress than venous endothelium and its contribution to the control of blood pressure would be more important. This study has been started to clarify the details of mechanosensitive Ca^{2+} mobilization in aortic endothelial cells.

EFFECTS OF HYPOTONIC STRESS ON $[Ca^{2+}]_i$

When hypotonic Krebs solution (-20%) was perfused over the cell, steep transient Ca^{2+} spikes were observed in almost all cells both in Ca^{2+}-free and Ca^{2+}-containing solutions (Fig.1A), suggesting that Ca^{2+} was released from intracellular store sites. The time course was much faster than those observed in endothelium from human umbilical vein (6), but similar to IP_3-mediated Ca^{2+} transients induced by agonists such as ATP (5). We then examined the cellular mechanisms by which Ca^{2+} was released from intracellular store sites. When hypotonic stress was applied to the cell pretreated with a phospholipase C inhibitor, neomycin, Ca^{2+} transient was not observed. Furthermore, bath application of the P_2 receptor antagonist, suramin, also inhibited the hypotonic stress-induced Ca^{2+} transient. These indicate that ATP was released from aortic endothelium in response to hypotonic stress, which then induced the Ca^{2+} transient. On the other hand, if the fast Ca^{2+} transient was inhibited by suramin or neomycin, hypotonic stress induced a gradual $[Ca^{2+}]_i$ elevation. Such a gradual $[Ca^{2+}]_i$ elevation was not residual activity of ATP, because a threshold concentration of ATP induced Ca^{2+} transients in an all-or-none manner, but not a gradual $[Ca^{2+}]_i$ elevation (Oike, unpublished observation). This indicates that aortic endothelium is capable of producing two types of Ca^{2+} response to hypotonic stress; i.e., ATP-mediated fast Ca^{2+} transient and slowly-developing $[Ca^{2+}]_i$ elevation. Exocytosis would be a possible mechanism by which ATP is secreted from endothelium in response to hypotonic stress. Cell swelling will unfold a creased membrane thereby making the fusion and rupture of ATP-containing vesicle, if present, easier. However, with such a mechanism, various substances would be released in a non-selective manner by cell swelling. Another mechanism would be the ATP release in relation to the activation of volume-sensitive Cl^- channel (Cl_{vol}). Hypotonic stress selectively activates Cl_{vol} in vascular endothelium and its amplitude is often in the nA range (3). Cl_{vol} admits not only Cl^- but also negatively charged molecules such as gluconate in the endothelium (4). ATP is also negatively charged, and the possibility of ATP permeation through Cl^- channel has been reported in CFTR Cl^- channel (8), though it is still controversial (9). However, in the case of Cl_{vol}, its pore size may not be enough to admit ATP because extracellular ATP induces open-channel block of Cl_{vol} (7). So the direct permeation of ATP through Cl_{vol} is unlikely. Many reports have suggested that P-glycoprotein, which is capable of releasing ATP, is not Cl_{vol} itself but may regulate Cl_{vol} (7). So cell swelling may activate a transporter such as P-glycoprotein thereby releasing ATP either dependently or independently of the activation of Cl_{vol}. Further investigation is needed to elucidate the pathway for ATP release by cell swelling in aortic endothelial cells.

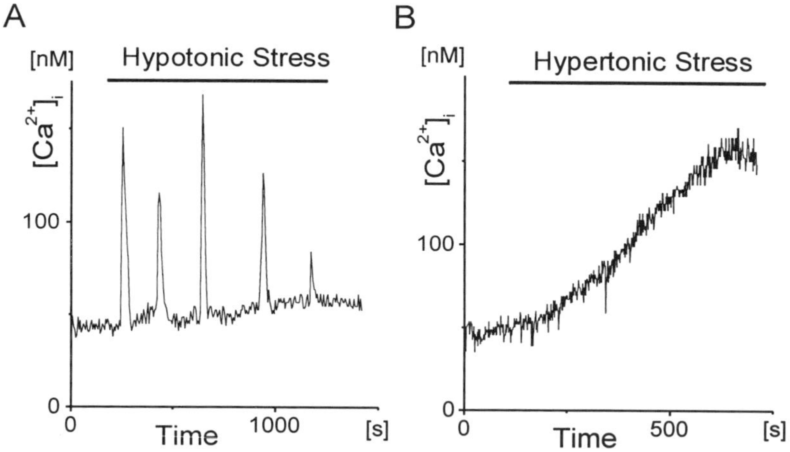

Figure 1. Effect of osmotic stress on intracellular Ca^{2+} concentration ($[Ca^{2+}]_i$) in bovine aortic endothelial cells. **A**. Hypotonic Krebs solution (-20%) induced repetitive Ca^{2+} spikes. **B**. Hypertonic Krebs solution (+20%) induced gradual increase in $[Ca^{2+}]_i$.

EFFECTS OF HYPERTONIC STRESS ON $[Ca^{2+}]_i$

On the other hand, when hypertonic solution (+20%) was applied to the cell, the increase of $[Ca^{2+}]_i$ was gradual in both Ca^{2+} free and Ca^{2+}-containing solutions

(Fig.1B), suggesting that hypertonic stress also mobilizes Ca^{2+} from intracellular store sites. The time course of Ca^{2+} release was similar as that elicited by hypotonic stress in endothelium from human umbilical cord vein, where $[Ca^{2+}]_i$ elevation was mediated by arachidonic acid (6). In the present experiment, we found that a phospholipase A_2 inhibitor, quinacrine, significantly inhibited the hypertonic stress-induced $[Ca^{2+}]_i$ elevation in bovine aortic endothelium. On the other hand, the metabolites of arachidonic acid were not involved in the hypertonic stress-induced Ca^{2+} transient because neither indomethacin, a cyclooxygenase inhibitor, nor NDGA, a lipoxygenase inhibitor, inhibited the $[Ca^{2+}]_i$ elevation. These results indicate that hypertonic stress-induced $[Ca^{2+}]_i$ elevation in aortic endothelium is mediated by arachidonic acid.

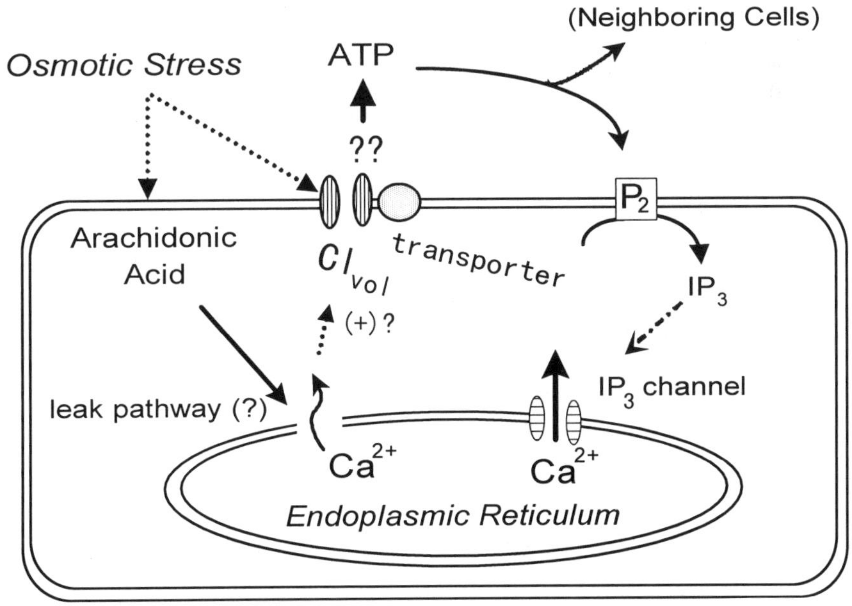

Figure 2. Schematic diagram of possible cellular mechanisms of Ca^{2+} release from intracellular store sites in response to osmotic stress in bovine aortic endothelial cells.

SUMMARY

As summarized in Fig.2, we have found a dual mechanisms of Ca^{2+} release in response to osmotic change in bovine aortic endothelial cells. Hypotonic stress induces the secretion of ATP from a cell, and Ca^{2+} is released from intracellular store sites by the activation of the P_2 receptor. However, when hypotonic stress-induced, ATP-mediated Ca^{2+} spikes are inhibited, hypotonic stress induces gradual release of Ca^{2+}. On the other hand, hypertonic stress can only induce a gradual $[Ca^{2+}]_i$ elevation, and the activation of phospholipase A_2 is probably involved in this mechanism. Previously we reported that arachidonic acid is responsible for the mechanosensitive Ca^{2+} transient in endothelium from human umbilical cord vein (6), but bovine aortic endothelium seems to possess an ATP-mediated mechanism as the osmosensitive machinery prior to one mediated by arachidonic acid.

As described above, vascular endothelium secretes various substances to control its local environment. Though the pathway of ATP release is still unknown, it is tempting to assume that it is related to the activation of Cl_{vol}, because it is the most important selective change in endothelium in response to hypotonic stress. It has been reported that the inhibition of Cl_{vol} attenuates endothelial proliferation (10). Furthermore, there is much evidence showing that Cl_{vol} admits large negatively charged cellular molecules (4). This may indicate the possibility of another important role of Cl_{vol} in endothelium; i.e., the involvement in the secretion of biologically active substances or cellular metabolites. Such a mechanism would bring good self-control of the local environment in response to osmotic stress in aortic endothelium.

ACKNOWLEDGEMENT

We thank Dr. Káte E. Creed for critical reading of the manuscript.

REFERENCES

1. Davis PF. Physiol Rev 1995; 75:519-560.
2. Milner P, Kirkpatrick KA, Ralevic V, Toothill V, et al. Proc R Soc Lond 1990; 241:245-248.
3. Nilius B, Eggermont Y, Voets T, Droogmans G. Gen Pharmacol 1996; 27:1131-1140.
4. Nilius B, Sehrer J, Droogmans G. Br J Pharmacol 1994;112:1049-1056.
5. Oike M, Ito Y. Eur J Pharmacol 1997; 319:291-298.
6. Oike M., Droogmans G, Nilius B. Proc Natl Acad Sci USA 1994; 91:2940-2944.
7. Okada Y. Am J Physiol 1997; 273:C755-C789.
8. Reisin IL, Prat AG, Abraham EH, Amara JF, et al. J Biol Chem 1994; 269:20584-20591.
9. Reddy MM, Quinton PM, Haws C, Wine JJ, et al. Science 1996; 271:1876-1879.
10. Voets T, Szucs G, Droogmans G, Nilius B. Pflügers Archiv 1995; 431:132-134.

Cell Volume Regulation: The Molecular Mechanism and Volume Sensing Machinery
Y. Okada, editor.

Ionic mechanisms of RVD in A6 cells: SA channel, $[Ca^{2+}]_i$, whole cell currents, and RVD

M. Sokabe, W. Yu and K. Takemoto

Department of Physiology, Nagoya University School of Medicine, Nagoya 466, Japan.

INTRODUCTION

Regulatory volume decrease (RVD) has been widely studied in recent years because of not only its physiological significance but its biological interest in the ingenious defensive system of the cell against mechanical stresses [1]. An increase in intracellular Ca^{2+} concentration ($[Ca^{2+}]_i$) could play a key signaling role in RVD. Upon hypotonic cell swelling, the increase in $[Ca^{2+}]_i$ eventually activates K^+ and Cl^- channels, which results in KCl efflux followed by water efflux and recovery of cell volume in a variety of cells [2]. The increase in $[Ca^{2+}]_i$ may be derived from both extracellular space and intracellular stores. It is generally assumed that the major source of the increase in $[Ca^{2+}]_i$ is extracellular space, because a dependence on extracellular Ca^{2+} of the ability to undergo RVD has been shown in many cell types, such as red blood cells, urinary bladder cells, intestinal epithelial cells and rabbit proximal tubules [2]. The question is which pathway is responsible for the hypotonically induced Ca^{2+} influx. One of the most probable routes which can mobilize Ca^{2+} from extracellular space is the Ca^{2+}-permeable, stretch-activated cation channel (SA-channel). It is suggested that membrane stretch upon hypotonic swelling induces Ca^{2+} influx through SA-channels, which leads RVD [3,4,5], nevertheless only a limited number of cell species has been reported to exhibit the $[Ca^{2+}]_i$ increase mediated by SA-channels during hypotonic stimulation [5,6]. The authors suggested the necessity of Ca^{2+} influx through SA-channels for the $[Ca^{2+}]_i$ increase and RVD, however, they did not show the net contribution of Ca^{2+} influx through SA-channels to the $[Ca^{2+}]_i$ increase and RVD.

The primary aim of the present study was to make it clear whether the Ca^{2+} influx through SA-channels activated by hypotonic cell swelling can solely induce RVD. A6 cell, an epithelial cell line of *Xenopus* kidney, may be a good model for investigating the involvement of Ca^{2+} influx through SA-channels in the hypotonically induced

$[Ca^{2+}]_i$ increase and RVD, because the expression of a SA-channel has already been confirmed in this cell type [7]. In addition RVD in A6 cells depends on extracellular Ca^{2+} and on coactivation of K^+ and Cl^- channels [8,9,10]. The second aim of this study was to elucidate the relationship between RVD, $[Ca^{2+}]_i$, and the conductances for K^+ and Cl^- during hypotonic shock in A6 cells. Our results suggested that a small increase in $[Ca^{2+}]_i$ derived from Ca^{2+} influx through SA-channels is enough to induce perfect RVD in A6 cells [11] and that a $[Ca^{2+}]_i$ induced-coactivation of K^+ and Cl^- conductances was indispensable for RVD [10].

INVOLVEMENT OF SA CHANNEL IN $[Ca^{2+}]_i$ INCREASE AND RVD

We at first analysed the relationship between changes in $[Ca^{2+}]_i$ and cell volume of A6 cells exposed to hypotonic solutions. $[Ca^{2+}]_i$ was evaluated, under a laser confocal scanning microscope system (Insight Plus, Meridian Instr Co. Ltd.), using a ratiometric measurement with two calcium sensitive fluorescent probes, fluo-3/AM and Fura Red/AM. These dyes preloaded in the cell were excited at 488 nm of argon laser line, and emissions were separated using a 530/30 nm band-pass and 605 nm long-pass filter set. Volume changes of individual cells during hypotonic stimulation were estimated from the height of vertical sections of cell images. Figure 1 shows a summaraized scheme of the changes in $[Ca^{2+}]_i$ and cell height under various experimental conditions.

Superfusion of A6 cells with hypotonic Ringer (55% osmolality; 60 NaCl, 3 KCl, 0.5 $CaCl_2$, 1 $MgCl_2$, 5.5 glucose, 1 Na-pyruvate, 10 HEPES, pH 7.4 adjusted by NaOH) induced cell swelling that occurred in the first two minutes after the stimulus onset followed by a recovery phase (Fig.1, line c), which was accompanied by a transient increase in $[Ca^{2+}]_i$. The $[Ca^{2+}]_i$ increase had a transient peak in less than one minute upon hypotonic stimulation and a small plateau during the stimulation (Fig.1, line a). To investigate the involvement of Ca^{2+} releases from intracellular stores in this RVD and $[Ca^{2+}]_i$ increase, cells were pretreated with 50 μ M thapsigargin. The pretreated cells maintained a small plateau of $[Ca^{2+}]_i$ without any detectable peak during hypotonic stimulation as shown in Fig.1 (line b). Thus, the early peak of $[Ca^{2+}]_i$ in control experiments is most likely derived from Ca^{2+} releases from intracellular stores. The height of the preteated cells changed in a similar way to that in control experiments (line c). Namely, RVD still occurred in such thapsigargin pretreated cells in spite of the complete absence of the peak in $[Ca^{2+}]_i$ (Fig. 1, line b) , and hence RVD can be inducible without participation of Ca^{2+} releases from intracellular stores.

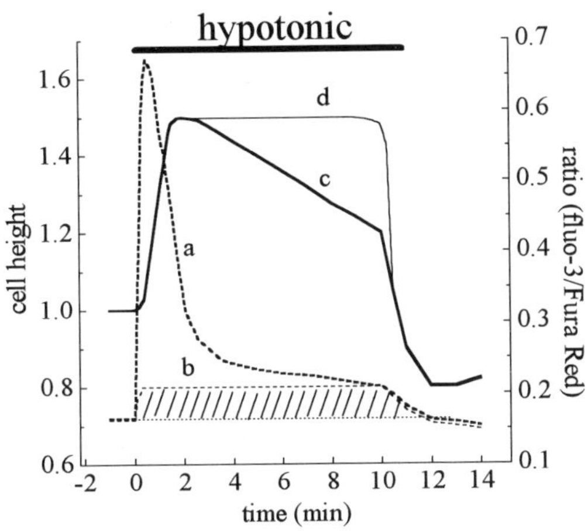

Figure 1. Schematic summary of changes in cell height and $[Ca^{2+}]_i$ during hypotonic stimulation. Relative cell height and $[Ca^{2+}]_i$ (ratio of fluorescence intensities) plotted against the time after the onset of hypotonic stimulation under various conditions. a, $[Ca^{2+}]_i$ increase under normal condition. $[Ca^{2+}]_i$ increases to a peak in less than one minute followed by a small plateau during the stimulation. b, $[Ca^{2+}]_i$ incease devoid of the peak in store depleted cells. c, Cell height changes showing RVD under normal condition. Essentially similar results could be obtained with either store depleted or external Ca^{2+}-deprived cells. d, Cell height changes without RVD in the store depleted cells exposed to Ca^{2+}-free or Gd^{3+} contained hypotonic solutions. Exposure to hypotonic Ringer (55% osmolality) was done from 0 min to 10 min at room temperature in all plots.

In contrast to the thapsigargin pretreatment, hypotonic stimulations with Ca^{2+}-free media caused an early peak of $[Ca^{2+}]_i$ increase but the later plateau was not observed (open area under line a in Fig.1), suggesting that the plateau is derived from Ca^{2+} entry across the plasma membrane. Although changes in cell height was relatively slow and small compared to control, RVD was still observed under this condition. Neither extracellular Ca^{2+} removal nor depletion of Ca^{2+} from intracellular stores could inhibit

RVD in A6 cells. We therefore tried to block both sources of the $[Ca^{2+}]_i$ increase simultaneously, that is cells pretreated with thapsigargin were exposed to Ca^{2+}-free hypotonic Ringer. Under this condition, $[Ca^{2+}]_i$ did not increase but rather slightly decreased, which might reflect the dilution of intracellular Ca^{2+} by hypotonic cell swelling. No RVD was observed under this condition (Fig. 1, line d). Taken together, Ca^{2+} influx across the plasma membrane seems to be primarily essential to induce RVD in A6 cells.

One of the most possible pathways of the Ca^{2+} influx is a Ca^{2+}-permeable SA-channel. To examine the involvement of SA-channels in RVD and Ca^{2+} influx, thapsigargin treated cells were swelled in hypotonic Ringer containing 10 μM Gd^{3+}, a potent blocker for the SA channel [7,12]. The cells showed no $[Ca^{2+}]_i$ increase and no RVD to hypotonic stimulation. However, as Gd^{3+} is known to block some types of voltage-dependent calcium channels [13,14], we examined the involvement of such voltage-dependent Ca^{2+} channels in the hypotonically induced $[Ca^{2+}]_i$ increase. Administration of 20 μM nifedipine caused a slight decrease in resting $[Ca^{2+}]_i$ but did not affect the increase in $[Ca^{2+}]_i$ during hypotonic stimulation. Administration of 50 μM verapamil caused a similar result. Moreover, we confirmed the absence of voltage-dependent calcium influx in A6 cells because depolarization by increasing extracellular K^+ concentration to 50 mM could not induce any increase in $[Ca^{2+}]_i$. We hence concluded that the inhibition of Ca^{2+} influx by Gd^{3+} during hypotonic stimulation was not caused by the blockade of voltage-dependent Ca^{2+} channels. Thus, it is highly likely that Gd^{3+} inhibited the hypotonically induced increase in $[Ca^{2+}]_i$ by blocking SA-channels in A6 cells. It is suggested that cell swelling by hypotonic Ringer induces Ca^{2+} influx through activated SA-channels leading a perfect RVD in store-depleted A6 cells.

WHOLE CELL CURRENTS AND CELL VOLUME

Recently it has been demonstrated by Nilius et. al. (1995) that hypotonically induced RVD of A6 cells is blocked by a variety of K^+ and Cl^- channel blockers, and that both K^+ and Cl^- channels are coactivated during hypotonic stimulation [9]. However, they studied the hypotonically induced conductances and volume regulation with separate preparations, whole cell clamped cells and intact cells, respectively, and

$[Ca^{2+}]_i$ changes during hypotonic stimulation and their role in RVD were not examined. Therefore, we measured hypotonically induced whole cell currents and either cell volume or $[Ca^{2+}]_i$ changes simultaneously in the same cells to analyze: 1) the characteristics of hypotonically induced whole cell currents and their role in RVD; and 2) the relationship between hypotonically induced-currents and -$[Ca^{2+}]_i$ increases.

In this study we used 80% hypotonic solutions (66.4mM mannitol was added to the above mentioned 55% hypotonic solution); this is because stronger stimulations such as employed in other studies (55% in [10], 60% in [9]) deteriorated the seal of whole cell clamped A6 cells preventing us from making stable current recordings. Although the volume- and $[Ca^{2+}]_i$ -responses in intact A6 cells subjected to 80% solutions were weak and slow compared to those reported by others [9, 10], basic properties of the responses including RVD (see Fig. 2C, △) were qualitatively the same. Actually our A6 cells which were exposed to stronger stimulations responded in a similar way to others'. Thus the weak and slow response in the present study arose simply from the weaker stimulation. More interestingly, as described below, those responses were enhanced in whole cell clamped cells even with such a relatively weak (80%) stimulation.

The effects of 80% hypotonic solution on an A6 cell volume under the whole cell clamped configuration are shown in Fig. 2A, where phase contrast images of a cell before, 4 min and 12 min after a hypotonic stimulation are presented. When exposed to a hypotonic solution, the cell started to swell rapidly (within a minute) and most cells reached their maximum volume within 5 min after hypotonic stimulation, followed by a recovery phase (Fig. 2C (c1)). What was different from others' reports was that in our experiment the cell volume oscillated during hypotonic stimulation (for a possible mechanism, see the following section). The fraction of the difference between the first peak and the well in the cell surface area with respect to the peak (% RVD) was used as an index for the extent of RVD, as shown in Fig. 2C (c1). Its mean value was approximately 92 percnt. Simultaneously recorded whole cell currents from the same cell are shown in Fig. 2B. The A6 cell, under the whole-cell voltage clamp configuration with a 125mM KCl solution in the pipette, exhibited current increases when exposed to hypotonic solution as shown in Fig. 2B (b2). The I-V curves corresponding to the steady state currents are summarized in Fig. 2B (b3). The currents showed time- and voltage-dependencies during strong depolarization, while no or little time- and voltage-dependencies over the range of -120 mV to 40 mV.

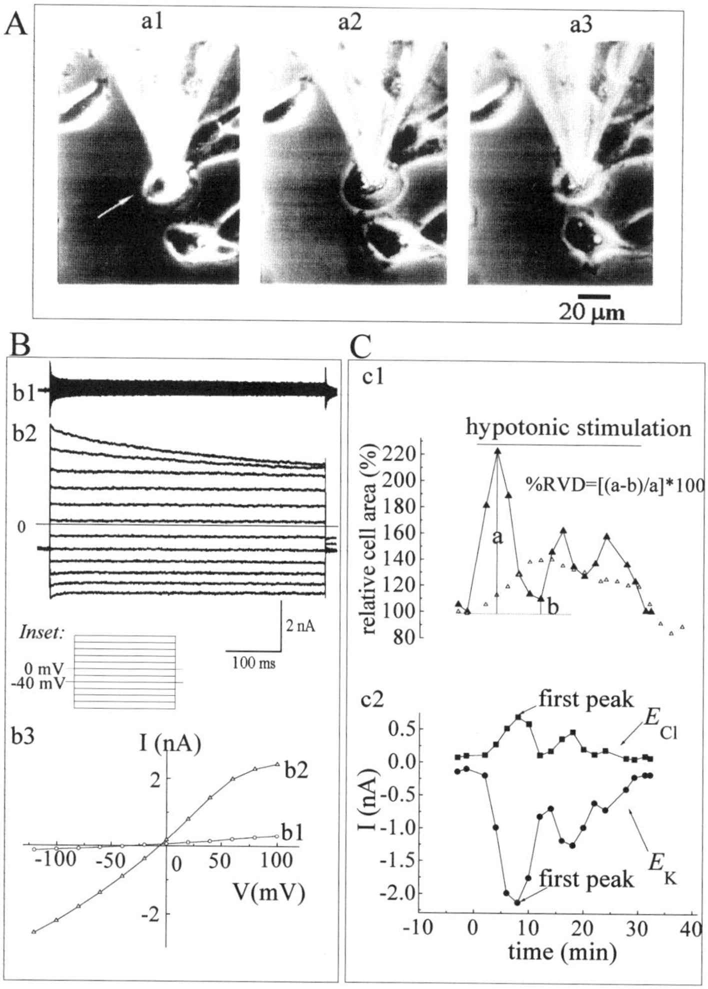

Figure 2. Hypotonically induced volume and whole cell current changes in an A6 cell. A. Phase contrast images of a whole cell clamped cell (indicated by arrow) before (a1), 4 min (a2) and 12 min (a3) after a hypotonic stimulation (80%). B. Whole cell current recordings before (b1) and 8 min (b2) after hypotonic stimulation and corresponding I-V curves (b3). *Inset*: voltage pulse protocol. C. Time courses of changes in: c1, cell surface area of a whole-cell clamp cell (▲) and intact cell (△); c2, outward currents at E_{Cl} (16.5 mV) and inward currents at E_K (-95.8 mV) during hypotonic stimulation. Horizontal bar indicates the period of hypotonic stimulation. Percent (%) RVD, as an index of the extent of RVD, was defined as the fraction of the difference between the first peak and the well in the cell area with respect to the peak .[10]

Fig. 2C (c2) shows time courses of steady-state currents during hypotonic stimulation. The outward currents were measured from I-V curves at the Cl^- equilibrium potential (E_{Cl}=16.5 mV) and inward currents at the K^+ equilibrium potential (E_K= -95.8 mV). Both of the two currents oscillated in a similar manner to the volume changes with a little delay (currents reached their first peak at 9 min after exposed to hypotonic solution, while for the volume of the same cells needed only 5 min). In this study, hypotonically induced increases (in volume, currents and $[Ca^{2+}]_i$) refer to those of the first peak with respect to the control level.

CHARACTERISTICS OF HYPOTONICALLY INDUCED CURRENTS

In order to dissect the hypotonically induced whole cell currents into defined ionic components, ion replacement experiments were done. Replacement of Na^+ with choline$^+$ showed little effect on the currents and RVD, indicating that Na^+ did not contribute to the hypotonically induced currents and RVD. As the reversal potentials of whole cell currents (-21.mV in isotonic solution; -0.3 mV in hypotonic solution, Fig. 2B(b3)) were between the equilibrium potentials of K^+ and Cl^-, the currents seemed to be carried by mainly K^+ and Cl^-, as has been suggested in a recent report [9]. The permeability ratios of Cl^- to K^+ (P_{Cl}/P_K) calculated from the mean reversal potential (E_m) according to Goldman-Hodgkin-Katz equation was 0.533 and 1.997 before and during hypotonic stimulation, respectively. In order to separate the two currents, K^+ was replaced with Cs^+, or Cl^- with glutamate. Replacement of K^+ with Cs^+ shifted the reversal potential closely to the Cl^- equilibrium potential, and replacement of Cl^- with glutamine shifted the reversal potential closely to the K^+ equilibrium potential, suggesting that the hypotonically induced whole cell currents were mainly composed of K^+ and Cl^- currents. However, what is unexpected was that replacement of K^+ or Cl^- not only inhibited the K^+ (at E_{Cl}) or Cl^- (at E_K) currents and shifted the reversal potential, but also inhibited the other current component. Replacement of K^+ with Cs^+ inhibited both K^+ and Cl^- currents, and similarly, replacement of Cl^- with glutamate inhibited both currents. What should be emphasized here is that when K^+ was replaced with Cs^+ or Cl^- with glutamate, cells swelled to a similar extent with a similar time course to the control, but RVD did not occur.

In order to elucidate whether the presence of Cl^- and K^+ was important to activate hypotonically induced K^+ and Cl^- currents, we studied the effects of channel blockers on the two currents in the presence of standard concentration of K^+ and Cl^-. It was found that K^+ channel blockers, TEA (10 mM) and quinine (500 μM), inhibited both of the hypotonically induced K^+ and Cl^- currents. Similarly, Cl^- channel blockers, NPPB (50 μM) and SITS (50 μM), inhibited hypotonically induced K^+ currents as well as the Cl^- currents. These results indicate that no matter how the currents were inhibited, K^+ and Cl^- currents could not be isolated from each other. When K^+ currents were blocked, Cl^- currents were also blocked, and *vice versa*, suggesting that there is a strong coupling between hypotonically induced K^+ and Cl^- currents. A strong linear relationship between hypotonically induced K^+ and Cl^- currents is demonstrated in Fig. 3. The hypotonically induced K^+ and Cl^- currents seem to depend on each other, inhibition of one leads an inactivation of the other. Similar results have been reported in human epididymal cells [15].

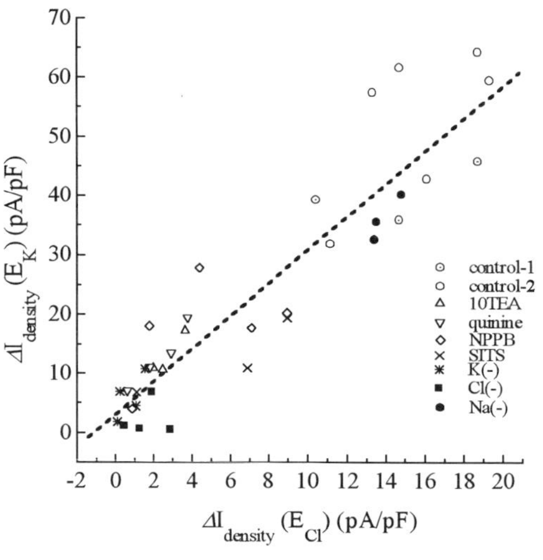

Figure 3. Relationship between hypotonically induced K^+ and Cl^- currents. Linear regression plot of K^+ currents *vs.* Cl^- currents (r = 0.92, P<0.001). Data were collected from various experiments as indicated in inset: ⊙, ○, control; △, 10 mM TEA; ▽, 500 μM quinine; ◇, 50 μM NPPB; ×, 50 μM SITS; ∗, $K^+ \rightarrow Cs^+$; ■, $Cl^- \rightarrow$ glutamate; ●, $Na^+ \rightarrow$ choline ($\rightarrow$ indicates ion replacement) [10].

RELATIONSHIP BETWEEN WHOLE CELL CURRENTS, $[Ca^{2+}]_i$, AND RVD

Simultaneous recordings of two dimensional cell surface areas and whole cell currents showed that: 1) hypotonically induced currents and volume changes shared a similar time course, where volume changes preceded current changes (Fig. 2C); 2) RVD was inhibited when the currents were inhibited irrespectively of the way of current inhibition. Relationship of the hypotonically induced Cl^- currents and %RVD is shown in Fig. 4. A very similar relationship between K^+ currents and %RVD was also found. The correlation coefficients for Cl^- and K^+ currents with RVD were 0.899 and 0.901, respectively. This result suggests that activation of the whole cell currents is essentially important for RVD.

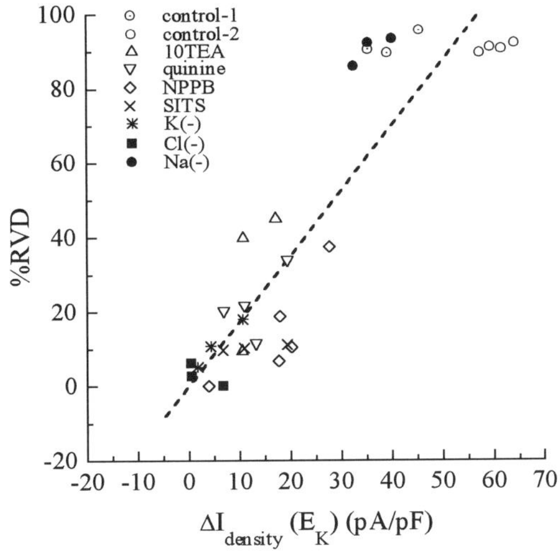

Figure 4. Relationship between whole cell (Cl^-) currents and RVD. Dashed line indicates a linear regression plot of Cl^- currents at E_K *vs.* %RVD ($r = 0.90$, $P<0.05$). Definition of %RVD can be seen in Fig. 2A and its legend. Symbols have the same meanings as in Fig.3 [10].

We measured $[Ca^{2+}]_i$ changes while recording whole cell currents in the same cells. When the cell was exposed to hypotonic solution, $[Ca^{2+}]_i$ increased to a peak value and then decreased to some extent and rose again; behaving in an oscillatory manner as observed for the volume and currents. What should be noted is that the time course of the $[Ca^{2+}]_i$ changes delayed a little to that of volume changes but advanced to the current changes, suggesting that the $[Ca^{2+}]_i$ increase induced the current increase. There is a good correlation between the $[Ca^{2+}]_i$ increases and current increases as shown in Fig. 5.

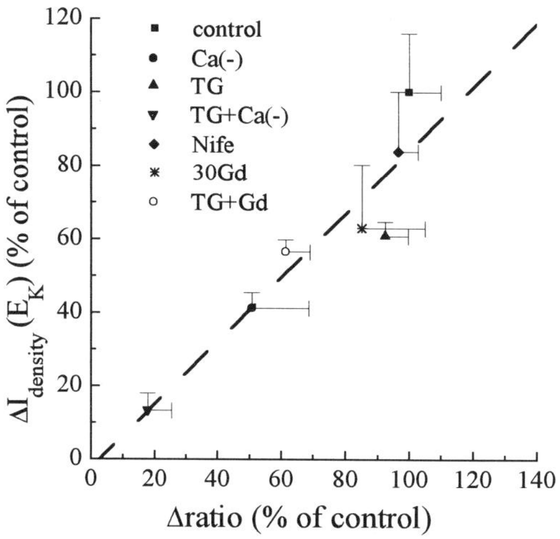

Figure-5. Relationship between whole cell (Cl^-) currents and $[Ca^{2+}]_i$. Linear regression plot of Cl^- currents at E_K *vs.* % Δratio ($r = 0.93$, $P<0.02$). ■, control; ●, external Ca^{2+}-free; ▲, thapsigalgin treatment; ▼, thapsigalgin + Ca^{2+}-free; ◆, 10 μM nifedipine; *, 30 μM $GdCl_3$; ○, thapsigalgin + $GdCl_3$.

Actually when the $[Ca^{2+}]_i$ increase was inhibited (*e.g.* by combined administration of thapsigargin and removal of extracellular Ca^{2+}), the K^+ and Cl^- currents were also significantly inhibited. To further ensure that the hypotonically induced currents depend on $[Ca^{2+}]_i$, we examined the effect of K^+ and Cl^- channel blockers on hypotonically induced $[Ca^{2+}]_i$ increases. A K^+ channel blocker (10 mM TEA) or a Cl^- channel blocker (50 μM NPPB) inhibited the currents increases but not $[Ca^{2+}]_i$ increases. The result indicates two fold. First, the blockers did not interfere the hypotonically activated Ca^{2+} mobilizing mechanisms including SA channel activation. Second, hypotonically induced K^+ and Cl^- currents were located down stream of the $[Ca^{2+}]_i$ increase. Thus a suggested signaling cascade is as follows: ***hypotonic shock → cell swelling → SA channel activation (and Ca^{2+} releases from internal stores) →$[Ca^{2+}]_i$ increases →coactivation of K^+ and Cl^- channels → K^+, Cl^- effluxes → RVD.***

EFFECTS OF WHOLE CELL CLAMP: IS IT A GOOD WAY TO STUDY RVD?

An interesting phenomenon we noticed in the present study was that the whole cell voltage clamped cells responded more violently than their intact neighboring cells; they swelled faster (5 *vs.* 17 min for the time to the first peak) and greater (223 *vs.* 137 % for the peak area increase), and $[Ca^{2+}]_i$ increased to a greater extent (0.134 *vs.* 0.042 for the fluorescence intensity ratio). One possible mechanism to induce violent response in the clamped cells may be that the membrane potential of the clamped cells during hypotonic shock is artificially clamped at the holding potential of -40 mV, which may be significantly lower compared with that in their intact neighbors. As the reversal potential during hypotonic shock was close to 0 mV (Fig. 2B (b3)) while that in isotonic solution was approximately -30 mV, the intact cells would be depolarized up to 0 mV during hypotonic shock. A similar result has been reported in A6 cells that the membrane potential is depolarized during hypotonic stimulation [9]. To hold the membrane potential at -40 mV would maintain a relatively stronger driving force for Ca^{2+} influx, and might cause more $[Ca^{2+}]_i$ increases in the clamped cells than their intact neighbors. To test this hypothesis, we changed the holding potential from -40 mV to +40 mV, however the hypotonically induced $[Ca^{2+}]_i$ increases did not change significantly. Therefore the difference of the driving force for Ca^{2+} may not be the major cause for the different behaviors between the clamped cells and intact cells.

The second possible mechanism may be the increasing pressure on the free dome of the cell, which was caused by the electrode placed on it during cell swelling. The pressure on the dome of the cell may affect in two ways: one is to make the cells swell relatively in a horizontal direction compared with their neighbors, thereby the clamped cells' surface area changed more violently than their neighbors (see Fig. 2A); another effect may be a stronger strain in the cell membrane, hence to activate SA channels to a greater extent which would cause greater increases in $[Ca^{2+}]_i$.

The third possible mechanism may be the diffusion between the cell and the pipette solutions in the clamped cells. It is believed that in the whole-cell clamp configuration, the intracellular ion concentration is "clamped" at that in the pipette solution due to continuous perfusion of intracellular space with pipette solution, thus RVD would not occur [9]. However, as shown in the present study, apparent RVD and $[Ca^{2+}]_i$ changes were observed in whole cell clamped cells. This suggested that the rate of perfusion of the cell with pipette solution was not fast enough to overcome the solutes and water exit during RVD. In order to examine the diffusion rate of solutions, 7/8 of KCl (123 mM) in the pipette was replaced with mannitol (final 16 mM KCl in the pipette). Under current clamp configuration, it took 1-2 min after the rupture of patch membrane to make the membrane potential attain its stable state. This result suggested that the intracellular ion concentrations were not "clamped" completely and immediately during hypotonic shock. Two antagonistic events would happen in the cell during hypotonic stimulation, solute (KCl) entry from the pipette and solute exit from the cell, and the rate of the latter may become to exceed at the time of the peak volume resulting in RVD. More systematic interpretation of the volume changes in the whole cell clamped cells during hypotonic stimulation is discussed as follows.

The volume change of the cell is determined by the net amount of water flux across the cell membrane and the pipette tip. In the whole cell clamped cells, there are three kinds of water flux: influx to the cytoplasm (f_i), effluxes to the bath (f_{eb}) and to the pipette (f_{ep}). Water influx from the pipette to the cell may be ignored since osmolality of the pipette solution would always be larger than that of the intracellular solution. In isotonic solution, as $f_i = f_{eb} + f_{ep}$, the cell keeps a constant volume. In an early stage of hypotonic stimulation, f_i increases, while f_{eb} decreases due to the osmotic gradient across the cell membrane. Although f_{ep} also increases, since Δf_i (Δ indicates the difference from the control level) would be much larger than Δf_{ep}, $f_i > f_{eb} + f_{ep}$, thereby cell swells during this stage. At the same time, the increase of f_{ep} (and the influx of ions into the cell from the pipette) contributes to keep the intracellular ion concentration higher than that of the intact cells. The higher intracellular concentration of solutes in the clamped cells would cause more water influx into the cell than intact cells leading a violent cell swelling. This enhanced swelling would strongly activate Ca^{2+} mobilizing mechanisms that lead activation of K^+ and Cl^- channels. With the increases of K^+ and Cl^- currents, f_{eb} increases while f_i and f_{ep} may change a little due to a partial perfusion of intracellular space. When $f_i < f_{eb} + f_{ep}$, cell volume starts to decrease, resulting in RVD. In the clamped cells, the intracellular osmolality increases again due to an inactivation of K^+ and Cl^- conductances and a constant diffusion of ions from the pipette, thereby the cell swells again. Such a mechanism may explain the observed oscillation of the volume, current and $[Ca^{2+}]_i$ in the present study.

A similar interpretation may stand in the $[Ca^{2+}]_i$ measurement during hypotonic stimulation. We measured the fluorescence ratio at the cell iterior, pipette tip and pipette shank (15 μm apart from the tip) using the pipette preloaded with the dyes. The fluorescent intensity ratio at the cell and the pipette tip changed in response to hypotonic stimulation, while that at the pipette shank unchanged, suggesting that the solute concentration of the cell interior was only partially clamped. In conclusion, although the whole cell clamping makes the process of RVD a bit complicated, it would be a useful method to investigate the volume regulation mechanism at least in A6 cells.

CONCLUSIONS

The signaling mechanism of regulatory volume decrease (RVD) could be decomposed into four steps, sensing and transducing, second messenger(s), intermediate signaling, and activation of solute efflux mechanisms. Using A6 cells we have revealed that a Ca^{2+}-permeable SA channel transduces the hypotonically induced cell swelling into a $[Ca^{2+}]_i$ increase. This $[Ca^{2+}]_i$ increase then activates interdependent K^+ and Cl^- channels *via* unknown signaling mechanisms, leading to RVD. However, there remain several important problems that should be clarified: 1) the intermediate mechanisms how the $[Ca^{2+}]_i$ increase activates the ion channels, 2) the mechanisms of hypotonically induced Ca^{2+} releases from internal stores, 3) single channel analyses of the hypotonically activated K^+ and Cl^- channels, and finally, 4) the mechanism of the interdependency between the K^+ and Cl^- channels.

ACKOWLEDGEMENT

This work was partly supported by a Grant-in Aid for Scientific Research (09257220) from the Ministry of Education, Science, Sports and Culture of Japan, and a grant from CREST of Japan Science and Tecnology Corporation.

REFERENCES

1. **Parker, J. C.** In defense of cell volume? *Am. J. Physiol.* 265: C1191- C1200, 1993.
2. **McCarty, N. A. and R. G. O'Neil.** Calcium signaling and cell volume regulation. *Physiol. Rev.* 72: 1037-1061, 1992.
3. **Christensen, O.** Mediation of cell volume regulation by Ca^{2+} influx through stretch-

activated channels. *Nature* 330: 66-68, 1987.

4. **Filipovic, D. and H. Sackin.** A calcium-permeable stretch-activated cation channel in renal proximal tubule. *Am. J. Physiol.* 260 : F119-F129, 1991.
5. **Okada, Y., A. Hazama, and W. L. Yuan.** Stretch-induced activation of Ca^{2+}-permeable ion channels is involved in the volume regulation of hypotonically swollen epithelial cells. *Neurosci. Res. Suppl.* 12: S5-S13, 1990.
6. **Chen, Y., S. M. Simasko, J. Niggel, W. J. Sigurdson, and F. Sacks.** Ca^{2+} uptake in GH_3 cells during hypotonic swelling; evidence for the sensory role of stretch activated ion channels. *Am. J. Physiol.* 270:C1790-1798, 1996.
7. **Kawahara, K. and K. Matsuzaki.** A stretch-activated cation channel in the apical membrane of A6 cells. *Jpn. J. Physiol.* 43: 817-832, 1993.
8. **Ehrenfeld, J., C. Raschi, and E. Brochiero.** Basolateral potassium membrane permeability of A6 cells and cell volume regulation. *J. Membr. Biol.* 138: 181-195, 1994.
9. **Nilius, B., J. Sehrer, P. De Smet, W. V. Driessche, and G. Droogmans.** Volume regulation in a toad epithelial cell line: role of coactivation of K^+ and Cl^- channels. *J. Physiol.* 487: 367-378, 1995.
10. **Yu, W. and M. Sokabe.** Hypotonically induced whole cell currents in A6 cells: relationship with cell volume and cytoplasmic Ca^{2+}. *Jpn. J. Physiol.* 47: 553-565, 1997
11. **Sokabe, M., K. Takemoto and W. Yu.** Involvement of SA-channels in the volume regulation to hypotonic stimulation in A6 cells. *Biophys. J.* 72: A399 (1997) (paper submitted).
12. **Yang, X. C. and F. Sachs.** Block of stretch-activated ion channels in Xenopus oocytes by gadolinium and calcium ions. *Science*, 243: 1068-1071, 1989.
13. **Biagi, B. A. and J. J. Enyeart.** Gadolinium blocks low- and high-threshold calcium currents in pituitary cells. *Am. J. Physiol.* 259 (*Cell Physiol.* 28): C515-C520, 1990.
14. **Docherty, R. J.** Gadolinium selectively blocks a component of calcium current in rodent neuroblastoma × glioma hybrid (NG108-15) cells. *J. Physiol.* 398: 33-47, 1988.
15. **Chan, H. C., W. O. Fu, Y. W. Chung, S. J. Huang, P. S. Chan, and P. Y. D. Wong.** Swelling-induced anion and cation conductances in human epididymal cells. *J. Physiol.* 478: 449-460, 1994.

Cell Volume Regulation: The Molecular Mechanism and Volume Sensing Machinery
Y. Okada, editor.

Cause and roles of mechanical stress in mammary gland

Kishio Furuya[a], Haruo Nakano[b] and Koh-ichi Enomoto[c]

[a]Department of Applied Biology, Kyoto Institute of Technology,
Kyoto, 606-8585, Japan

[b]Department of Cell Physiology, National Institute for Physiological Sciences,
Okazaki, 444-8585, Japan

[c]Department of Physiology, Shimane Medical University,
Izumo, 693-8501, Japan

SUMMARY

In cultured mammary epithelial cells, mechanical stimulation induced Ca^{2+} wave that spread cell-to-cells with velocity of 5-10 μm/sec. The Ca^{2+} wave can propagate between separated cells or colonies, indicating the involvement of extracellular active substances. HPLC analysis of the supernatant of pipetting-induced stimulation of mammary tumor cell culture revealed the active substances to be ATP, UTP and UDP. The receptor of the nucleotides on the cells was found to be P_{2U} type purinoceptor using the cloning with RT-PCR and functional expression in K-562 cells. Using a newly developed co-culture of myoepithelial and secretory epithelial cells in mammary gland, mechanically induced Ca^{2+} wave was found to be triggered by myoepithelial contraction with oxytocin. Furthermore, myoepithelial cells were found to possess P_{2Y} purinergic receptors and respond to the Ca^{2+} wave. These results indicate that the Ca^{2+} wave is evoked by mechanical stress of myoepithelial contraction in mammary gland and mediates paracrine interaction between myo- and secretory epithelial cells forming positive feedback to expel milk.

INTRODUCTION

Intracellular Ca^{2+} plays essential roles in signaling in every type of cells. Even in non-excitable cells, stimulants induce Ca^{2+} signals that possess all-or-none like property sometimes and it spreads over the whole cell or transmits to the neighbor cells like an electrical signal in excitable cells. In some cells including mammary epithelial cells [1-4], airway epithelial cells [5], glial cells [6], leukemia cells [7] and so on, mechanical stimulation induced the intercellular Ca^{2+} waves. The Ca^{2+} waves spread with two different mechanisms. One is transmitting intracellular messengers (e.g. IP_3) through gap junctions between the cells. This case is observed in airway epithelium and glial cells in the brain [5, 8, 9]. The other mechanism is diffusing extracellular messenger (e.g., ATP) through the medium. Leukemia cells [7] and mammary epithelial cells [2, 3, 4, 10] were the case. Here, we summarized propagation mechanisms of mechanically induced Ca^{2+} waves in mammary epithelial cells.

In mammary gland, myoepithelial cells envelop alveoli of secretory epithelial cells and contract with oxytocin released from posterior pituitary. The contraction of myoepithelial cells compresses the alveoli and induces milk ejection. For an origin of mechanical stimulation in mammary gland, it is conceivable that the contraction of myoepithelial cells causes mechanical stress to the epithelial cells. In fact, the secretory epithelial cells were known to suffer the morphological deformation by myoepithelial contraction [11]. Here, we showed that the contraction of myoepithelial cells really induced Ca^{2+} wave in secretory epithelial cells in newly developed co-culture. We also discussed the roles of the Ca^{2+} wave or ejected nucleotides as an autocrine or paracrine mediator in mammary gland.

RESULTS AND DISSCUSSION

Mechanically induced Ca^{2+} wave

Mechanical stimulation of a mammary epithelial cell in culture evoked an increase of intracellular Ca^{2+} concentration ($Ca^{2+}{}_i$) and it spread to the surrounding cells like a wave (Fig 1A). The signal spread over a radius of about 200 μm with velocity of 5-10 μm/sec (Fig 1B). The Ca^{2+} wave spread not only within the colony but also to the separated adjacent colony (Fig 1A, [2], [4]). Further, the spreading pattern of Ca^{2+} wave was biased to the direction of the water flow. These findings indicated that the Ca^{2+} wave in mammary epithelial cells transmitted cell-to-cell by the diffusion of extracellular active substances ejected from the mechanically stimulated cell.

Release of nucleotides as active substances

An application of supernatant after pipetting of mammary tumor cells induced the increase of $Ca^{2+}{}_i$. So, the supernatant was considered to contain active substances. The activity was completely abolished by the treatment with phosphodiesterase or pyrophosphatase but not with heat, protease, ATPase, and DNase, suggesting the active substances to be pyrophospho-nucleotides. Using fractionation and HPLC analysis, the active substances were identified as ATP, UTP and UDP [10].

Involvement of P_{2U} purinergic receptor

Mammary epithelial cells responded to extracellularly applied nucleotides, such as ATP, UTP, ADP, UDP but not to AMP, UMP [3]. The potency to increase $Ca^{2+}{}_i$ was ATP, UTP > UDP, ADP >> AMP, UMP. Depletion of Ca^{2+} in solution did not affect the responses, suggesting that the Ca^{2+} source was intracellular stores [3]. Among the subtypes of purinergic receptor, these properties corresponded to those of P_{2U} type

($P2Y_2$ in recent nomenclature of purinergic receptors [12, 13]).

These findings implied that the nucleotides (ATP, UTP, UDP) were released from the mechanically stimulated cell, diffused, and activated P_{2U} receptor on the surrounding cells. Suramin, a blocker of P_2 type purinergic receptors, suppressed mechanically induced Ca^{2+} wave, supporting the idea [10].

A B

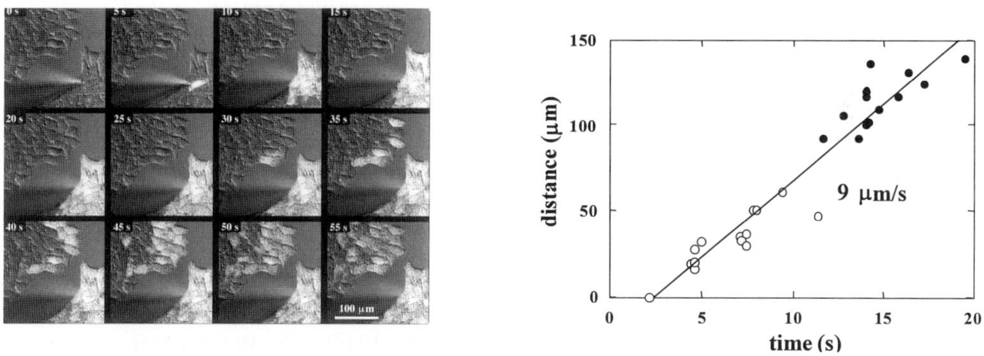

Figure 1. **A.** Mechanically induced Ca^{2+} wave in mammary epithelial cells. Slightly touching a mammary epithelial cell with a heat-blunted glass microelectrode induced $Ca^{2+}{}_i$ increase in the touched cell (at 5 s). The $Ca^{2+}{}_i$ increase spread to the surrounding cells like a wave (10 - 25 s). The signal further spread to a separated colony (30 s -), indicating involvement of released active substances. Cultured mammary tumor cells were loaded with indo-1 (AM) and observed in a confocal microscope. Nomarski image and ratio image of two fluorescence of indo-1 (F405 / F480) were superimposed. **B.** Relationships between distance from stimulated point and time lag of $Ca^{2+}{}_i$ increase after stimulation in each cell. $Ca^{2+}{}_i$ increase was measured using ratio of fura-2 fluorescence excited with two wavelength (F340 / F360) in mammary tumor cells in culture. The slope means velocity of the Ca^{2+} wave. Open and close circles show different colonies.

Cloning and expression of P_{2U} receptor

Using the cloned sequence of P_{2U} purinergic receptor in NG108-15 cells [14], P_{2U} receptor in mammary tumor cell was cloned by the RT-PCR [15]. Obtained PCR product was further transformed into expression vector and expressed in K-562 human leukemia cell line. Transfected K-562 cells responded to ATP, UTP and UDP as similar potency as mammary epithelial cells did. The cloned P2U receptor consisted of 373 amino acids and highly homologous with those of NG108-15 cells and human airway epithelium [16] (Fig 3A). Analysis of deduced amino acid sequence showed existence of 7 putative hydrophobic trans-membrane domains (Fig 3B). There were 2 consensus N-linked glycosylation sites and 4 residues highly conserved G-protein coupled receptors [15].

A

```
P2UMM    1: MAADLEPWNSTINGTWDGDELGYKCRFNEDFKYVLLPVSYGVVCVLGLCLNVVALYIFLC  60
P2UNG    1: MAADLEPWNSTINGTWEGDELGYKCRFNEDFKYVLLPVSYGVVCVLGLCLNVVALYIFLC  60
P2UHM    1: MAADLGPWNDTINGTWDGDELGYRCRFNEDFKYVLLPVSYGVVCVLGLCLNAVALYIFLC  60

P2UMM   61: RLKTWNASTTYMFHLAVSDSLYAASLPLLVYYYARGDHWPFSTVLCKLVRFLFYTNLYCR 120
P2UNG   61: RLKTWNASTTYMFHLAVSDSLYAASLPLLVYYYARGDHWPFSTVLCKLVRFLFYTNLYCS 120
P2UHM   61: RLKTWNASTTYMFHLAVSDALYAASLPLLVYYYARGDHWPFSTVLCKLVRFLFYTNLYCS 120

P2UMM  121: ILFLNCISVHRCLGVLRPLHSLRWGRARYARRVAAVVWVLVLACQAPVLYFVTTSVRGTR 180
P2UNG  121: ILFLTCISVHRCLGVLRPLHSLRWGRARYARRVAAVVWVLVLACQAPVLYFVTTSVRGTR 180
P2UHM  121: ILFLTCISVHRCLGVLRPLRSLRWGRARYARRVAGAVWVLVLACQAPVLYFVTTSARGGR 180

P2UMM  181: ITCHDTSARELFSHFMAYSSVMLGLLFAVPFSVILVCYVLMARRLLKPAYGTTGGLPRAK 240
P2UNG  181: ITCHDTSARELFSHFVAYSSVMLGLLFAVPFSVILVCYVLMARRLLKPAYGTTGGLPRAK 240
P2UHM  181: VTCHDTSAPELFSRFVAYSSVMLGLLFAVPFAVILVCYVLMARRLLKPAYGTSGGLPRAK 240

P2UMM  241: RKSVRTIALVLAVFALCFLPFHVTRTLYYSFRSLDLSCHTLNAINMAYKITRPLASANSC 300
P2UNG  241: RKSVRTIALVLAVFALCFLPFHVTRTLYYSFRSLDLSCHTLNAINMAYKITRPLASANSC 300
P2UHM  241: RKSVRTIAVVLAVFALCFLPFHVTRTLYYSFRSLDLSCHTLNAINMAYKVTRPLASANSC 300

P2UMM  301: LDPVLYFLAGQRLVRFARDAKPPTEPTPSPQARRKLGLHRPNRT-V-R-KDLSVSSDNSR 357
P2UNG  301: LDPVLYFLAGQRLVRFARDAKPPTEPTPSPQARRKLGLHRPNRT-V-R-KDLSVSSDDSR 357
P2UHM  301: LDPVLYFLAGQRLVRFARDAKPPTGPSPATPARRRLGLRRSDRTDMQRIGDVLGSSEDFR 360

P2UMM  358: RTESTPAGSE-TPYVRL                                            373
P2UNG  358: RTESTPAGSE-TKDIRL                                            373
P2UHM  361: RTESTPAGSENTKDIRL                                            377
```

B

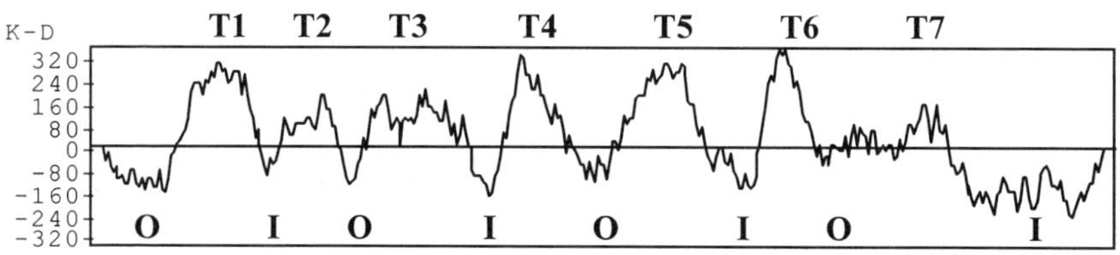

Figure 2. **A.** Deduced amino acid sequence of cloned P_{2U} receptor in mammary tumor cells (MM) and alignment to other P_{2U} receptors of NG108-15 (NG) and human epithelial cells (HM). **B.** Hydrophobicity plot of cloned P_{2U} receptor in mammary tumor cells. The plot is consistent with existence of seven putative trans-membrane domains (T1-7). I and O show intracellular and extracellular regions, respectively.

Induction of Ca^{2+} wave by myoepithelial contraction

Many types of tissues in the body including skin, blood vessel and airway accept mechanical stress and evoke Ca^{2+} wave [17]. A possible role of Ca^{2+} wave is to transmit a local stimulation to surrounding cells to coordinate activity in extend area of the tissue. In some cases including wound repair and antigen reaction enhancement, extracellular nucleotides share essential part of cell-to-cell signaling. In mammary gland, myoepithelial cells surrounding the alveoli contract by oxytocin released from posterior pituitary. The contraction of myoepithelial cells may exert mechanical stress on secretory epithelial cells and raise the Ca^{2+} wave.

We have developed a new culture method of mammary alveoli that contained considerable amount of myoepithelial as well as the secretory epithelial cells (Fig. 3). The application of oxytocin on the culture induced $Ca^{2+}{}_i$ increase and contraction of myoepithelial cells at first, and after some delays the contraction induced Ca^{2+} waves through the secretory epithelial cells. The Ca^{2+} wave diminished by suramin, an antagonist of P_2 purinergic receptors (Furuya K et al., in preparation).

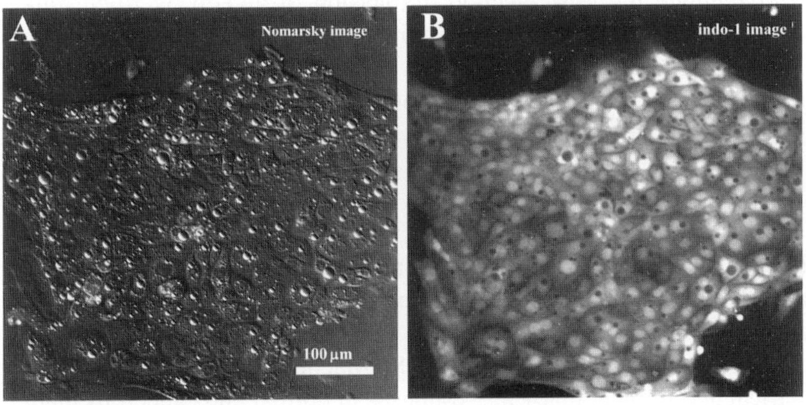

Figure 3. Co-culture of myo- and secretory epithelial cells in mouse mammary gland. **A;** Nomarsky (differential interference contrast) image, **B;** indo-1 fluorescence image excited with 364 nm observed with 480 nm, using a laser confocal microscope. Cells were cultured on collagen gel coated coverslips 2-4 days.

Involvement of P_{2Y} purinergic receptor in myoepithelial cells

Application of UTP (10 µM) on the co-culture induced $Ca^{2+}{}_i$ increase in secretory epithelial cells via activation of P_{2U} purinergic receptors, but not in myoepithelial cells. However, application of ATP (10 µM) induced $Ca^{2+}{}_i$ increase in both myoepithelial cells and secretory epithelial cells. Myoepithelial cells could contract by ATP. This finding suggested that the myoepithelial cells had another type of purinergic receptor. We have succeeded to isolate and culture the myoepithelial cells in mammary gland,

and investigated the character [18]. The potency of nucleotides to the $Ca^{2+}{}_i$ responses was ATP = ADP > UTP > UDP. $Ca^{2+}{}_i$ increase was induced even in a Ca^{2+}-free solution and inhibited by suramin. These findings implied the existence of a purinergic P_{2Y} type ($P2Y_1$ in the recent nomenclature [12, 13]) receptor in the myoepithelial cells. This result meant that myoepithelial cells could respond to the Ca^{2+} wave initiated by the contraction of myoepithelial cells.

Roles of Ca^{2+} wave and purinergic receptors in mammary gland

Our findings draw a new scheme of cell-to-cell interaction in mammary gland. The contraction of myoepithelial cells by oxytocin released from posterior pituitary functions as a mechanical stress to secretory epithelial and evokes the Ca^{2+} wave via nucleotide ejection and P_{2U} activation (autocrine effect). The Ca^{2+} waves or ejected nucleotides, in turn, induce $Ca^{2+}{}_i$ increase and contraction of myoepithelial cells (paracrine effect). So, myo- and secretory epithelial cells interact closely with both mechanical and purinergic couplings and they form a positive feedback system to enhance or regulate the milk ejection. Expansion of alveoli by filled milk may also give mechanical stress to the secretory epithelial cells. Of course, oxytocin is a prime factor in milk ejection reflex, but Ca^{2+} waves induced by these mechanical stresses may work as second factor to enhance, coordinate or regulate the milk ejection process.

RERERENCES

1. Furuya K, Enomoto K (1990) Real-time imaging of intracellular calcium change with simultaneous single channel recording in mammary epithelial cells. Brain Res Bull 25:779-781.
2. Enomoto K, Furuya K, Yamagishi S, Maeno T (1992) Mechanically induced electrical and intracellular Ca^{2+} responses in normal and cancerous mammary cells. Cell Calcium 13:501-511.
3. Furuya K, Enomoto K, Yamagishi S (1993) Spontaneous calcium oscillation and mechanically and chemically induced Ca^{2+} responses in mammary epithelial cells. Pflügers Archiv 422:295-304.
4. Furuya K, Enomoto K, Maeno T, Yamagishi S (1993) Mechanically induced calcium signal in mammary epithelial cells. Jpn J Physiol 43:S105-108.
5. Sanderson MJ, Charles AC, Dirksen ER (1990) Mechanical stimulation and intracellular communication increases intracellular Ca^{2+} in epithelial cells. Cell Regulation 1: 585-596.
6. Charles AC, Merrill JE, Dirksen ER, Sanderson MJ (1991) Intercellular signaling in glial cells: Calcium waves and oscillations in response to mechanical stimulation and glutamate. Neuron 6: 983-992.
7. Osipchuk Y, Cahalan M (1992) Cell-to-cell spread of calcium signals mediated by ATP receptors in mast cells. Nature 359: 241-244.

8. Charles AC, Christian CG, Zhu D, Kidder GM, Dirksen ER, Sanderson MJ (1992) Intercellular calcium signaling via gap junction in glioma cells. J. Cell Biol. 118: 195-201.
9. Boitano S, Dirksen ER, Sanderson MJ (1992) Intercellular propagation of calcium waves mediated by inositol trisphosphate. Science 258: 292-295.
10. Enomoto K, Furuya K, Yamagishi S, Oka T, Maeno T (1994) The increase in the intracellular Ca^{2+} concentration induced by mechanical stimulation is propagated via release of pyrophosphorylated nucleotides in mammary epithelial cells. Pflügers Archiv 427, 533-542.
11. Pitelka D, Hamamoto S, Duafala JG, Memanic MK (1973) Cell contacts in the mouse mammary gland. I. Normal gland in postnatal development and the secretory cycle. J. Cell Biol. 56: 797-818.
12. Fredholm BB, Abbracchio MP, Burnstock G, Daly JW, Harden TK, Jacobson KA, Leff P, Williams M (1994) Nomenclature and classification of purinoceptors. Pharmacological Reviews 46 :143-56.
13. Inoue K, Koizumi S, Ueno S (1996) Implication of ATP receptors in brain functions. Progress Neurobiol. 50: 483-492.
14. Lustig KD, Shiau AK, Brake AJ, Julius D (1993) Expression cloning of an ATP receptor from mouse neuroblastoma cells. Proc. Natl. Acad. Sci. 90: 5113-5117.
15. Enomoto K, Furuya K, Moore R, Yamagishi S, Oka T, Maeno T (1996) Expression cloning and signal transduction pathway of P_{2U} receptor in mammary tumor cell. Biological Signals 5: 9-21.
16. Parr CE, Sullivan DM, Paradiso AM, Lazarowski ER, et al. (1994) Cloning and expression of a human P_{2U} nucreotide receptor, a target for cystic fibrosis pharmacotherapy. Proc. Natl. Acad. Sci. 91: 3275-3279.
17. Sanderson MJ, Charles AC, Boitano S, Dirksen ER (1994) Mechanisms and function of intercellular calcium signaling. Mol. Cell. Endocri. 98: 173-187.
18. Nakano H, Furuya K, Furuya S, Yamagishi S (1997) Involvement of P2-purinergic receptor in intracellular calcium responses and the contraction of mammary myoepithelial cells. Pflügers Archiv 435:1-8

Cell Volume Regulation: The Molecular Mechanism and Volume Sensing Machinery
Y. Okada, editor.

ATP release from swollen or CFTR-expressing epithelial cells

Akihiro Hazama, Akiko Miwa, Takashi Miyoshi, Takahiro Shimizu, Yasunobu Okada

Department of Cellular and Molecular Physiology, National Institute for Physiological Sciences, Okazaki 444-8585, Japan

SUMMARY

Upon a hypotonic challenge most animal cells show swelling followed by regulatory volume decrease (RVD). Although this mechanism is mainly attained by the loss of KCl from the cell, not only ions but also amino acids and ATP are reported to exit the cell under hypotonic conditions. As ATP can modulate ion channels via purinergic receptors, RVD is possibly modulated by the extracelluar ATP. Actually recent reports suggest that ATP release is involved in the process of RVD. The pathway for ATP transport, however, is still unknown. We examined the ATP release from Intestine 407 cells under hypotonic conditions and got data suggesting that the pathway for ATP release is distinct from volume-sensitive Cl^- channel and CFTR. The ATP release assay from a CFTR-overexpressing cell line suggested that CFTR is not the pathway for ATP release but a modulator of ATP transport pathway.

IONIC MECHAMISM OF RVD AND ROLE OF ATP

Most animal cells can regulate their volume even under anisotonic conditions. We have studied the ionic mechanism of regulatory volume decrease (RVD) in human epithelial Intestine 407 cells. By measuring cell volume changes, membrane potential and ionic currents under two-microelectrode voltage-clamp, we have shown that the RVD is attained by KCl efflux due to activation of Ca^{2+}-dependent, quinine- and Ba^{2+}-sensitive K^+ conductance and Ca^{2+}-independent, SITS-sensitive Cl^- conductance [1]. In association with osmotic swelling, increases in cytosolic free Ca^{2+} were observed by means of Ca^{2+}-selective microelectrodes and fura-2 [2]. The Ca^{2+} response to hypotonic stress was found to consist of both Ca^{2+} influx and Ca^{2+} release [3]. Whole-cell recordings demonstrated that swelling-induced Ca^{2+} influx is largely due to membrane stretch-induced activation of Ca^{2+}-permeable cation channels [4]. Fura-2 studies showed that swelling-associated Ca^{2+} release is the result of Ca-induced Ca release from ryanodine-sensitive Ca stores [3]. Figure 1 represents the schematic model of ionic mechanism of RVD in Intestine 407 cells.

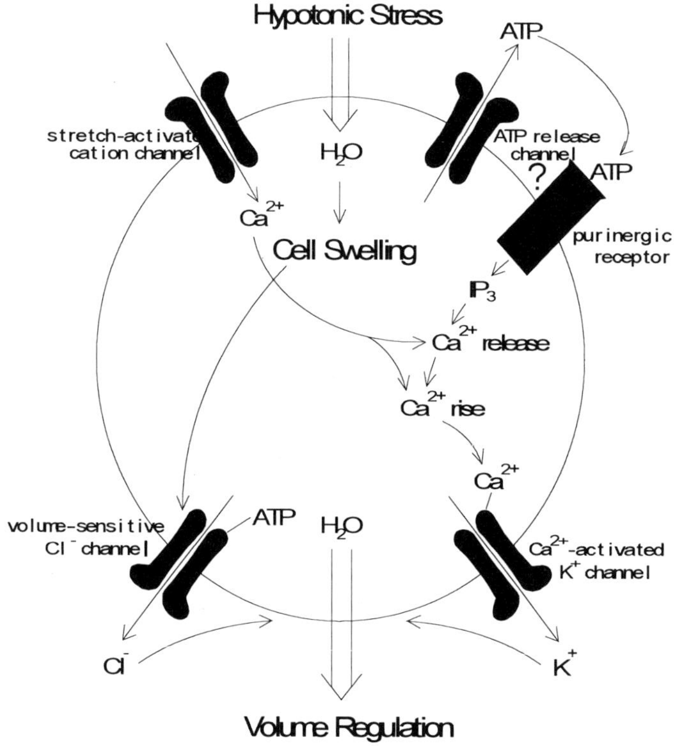

Figure 1. Schematic model of ionic mechanism of RVD in Intestinal 407 cells and our current working hypothesis of involvements of ATP release channel and ATP receptor.

Recently, it has been reported that ATP release from swollen cells is involved in RVD

[5,6]. Also, CFTR has been suggested to act as channels not only for Cl^- but also for ATP in a CFTR-overexpressing cell line [7]. Thus, we have started the studies to answer following questions: Whether ATP release is induced upon osmotic swelling in Intestine 407 cells, too? Whether CFTR is expressed in Intestine 407 cells? Whether CFTR expression enhances ATP release?

ATP RELEASE FROM SWOLLEN EPITHELIAL CELLS DEVOID OF CFTR EXPRESSION

To address the first question, we first measured the ATP release from Intestine 407 cells upon a hypotonic challenge. After hypotonic solution (60% hypotonicity) was applied to the cell, the concentration of ATP in the bath solution increased [8]. The viability of cells measured with MTT method did not change under hypotonic conditions. This indicates that ATP was released from the cell via a special pathway but not by a simple cell damage. To check the possibility that ATP can pass through volume-sensitive Cl^- channel, we applied SITS (0.5 mM) [8], which blocks volume-activated Cl^- channels in Intestine 407 cells [9]. SITS could not inhibit swelling-induced ATP release from Intestine 407 cells. Thus, it is suggested that ATP may pass the membrane via a pathway different from volume-sensitive Cl^- channel. As CFTR is also a candidate for the pathway of ATP, we examined the expression of CFTR mRNA in Intestine 407 cells using RT-PCR method. Figure 2 shows that no CFTR expression was detected in this cell line. Thus swelling-induced ATP releasing pathway is different from CFTR in Intestine 407 cells. Taken together, it appears that ATP pathway which is activated by osmotic swelling is distinct from the volume-sensitive Cl^- channel and CFTR Cl^- channel.

Intestine 407 cells are known to possess ATP receptors, the stimulation of which increases cytosolic Ca^{2+} [10]. Patch-clamp studies showed that extracellular ATP does not enhance volume-sensitive Cl^- currents [11]. Taken together, there is a possibility that swelling-induced ATP release enhances Ca^{2+}-activated K^+ channels via the ATP receptor, thereby facilitating RVD in Intestine 407 cells, as schematically depicted in Fig. 1.

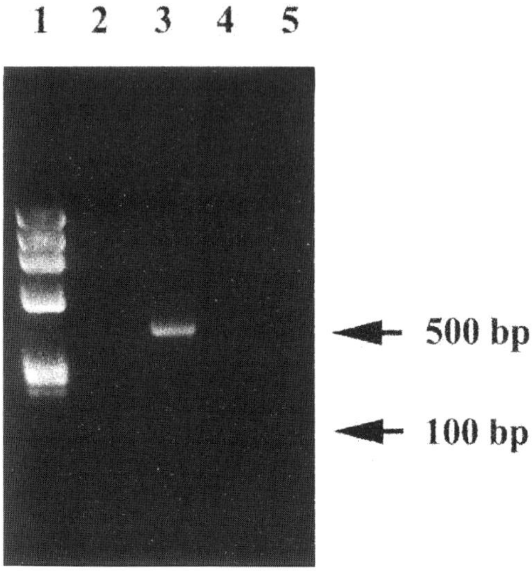

Figure 2. RT-PCR analysis of CFTR mRNA (lane-2) from Intestine 407 cells.
The first lane represents ϕX174 RF DNA/*Hae* III fragments as size markers. Lane-3 represents G3PDH as a positive control. Lanes-4 and -5 represent negative controls (without reverse transcriptase) for CFTR and G3PDH, respectively. The signal for CFTR was not observed in lane-2 (100 bp), while G3PDH was detected as shown in lane-3 (500 bp).

ATP RELEASE FROM CFTR-EXPRESSING EPITHELIAL CELLS

It was reported that ATP secretion is mediated by CFTR [12], although it is not the case of swelling-induced ATP release in Intestine 407 cells. To evaluate the role of CFTR in ATP secretion, we compared ATP secretion under hypotonic conditions from a CFTR-overexpressing cell line (C127/CFTR) and its control cell line (C127I). We observed that a hypotonic stress induced more marked increases in ATP secretion from C127/CFTR cells than from C127I [13]. This ATP release was not inhibited by glibenclamide which is a blocker not only of CFTR but also of volume-sensitive Cl^- channel [14]. This result suggests that CFTR can stimulate an ATP transport pathway which is distinct from CFTR Cl^- channel and volume-sensitive Cl^- channel. We assume that the pathway is an ATP-specific channel. Since CFTR is expressed in normal intestinal epithelial cells [15], enhancement of ATP release in anisotonic conditions by CFTR might help the cell to regulate their volume rapidly.

This work was supported by a Grant-in-Aid for Scientific Research on Priority Areas of "Channel-Transporter Correlation" from the Ministry of Education, Science, and Culture of Japan, and by CREST (Core Research for Evolutional Science and Technology) of Japan Science and Technology Corporation (JST).

REFERENCES

1 Hazama A, Okada Y. Ca^{2+} sensitivity of volume-regulatory K^+ and Cl^- channels in cultured human epithelial cells. J Physiol 1988; 402: 687-702.

2 Hazama A, Okada Y. Biphasic rises in cytosolic free Ca^{2+} in association with activation of K^+ and Cl^- conductance during the regulatory volume decrease in cultured human epithelial cells. Pflugers Arch 1990; 416: 710-714.

3 Hazama A, Okada Y. Involvement of Ca^{2+}-induced Ca^{2+} release in the volume regulation of human epithelial cells exposed to a hypotonic medium. Biochem Biophys Res Commun 1990; 167: 287-293.

4 Okada Y, Hazama A, Yuan W-L. Stretch-induced activation of Ca^{2+}-permeable ion channels is involved in the volume regulation of hypotonically swollen epithelial cells. Neurosci Res 1990; 12: S5-S13.

5 Wang Y, Roman R, Lidofsky SD, Fitz JG. Autocrine signaling through ATP represents a novel mechanism for cell volume regulation. Proc Natl Acad Sci USA 1996; 93: 12020-12025.

6 Roman RM, Wang Y, Lidofsky SD, Feranchak AP, Lomri N, Scharschmidt BF, Fitz JG. Hepatocellular ATP-binding cassette protein expression enhances ATP release and autocrine regulation of cell volume. J Biol Chem 1997; 272: 21970-

21976.

7 Reisin IL, Prat AG, Abraham EH, Amara JF, Gregory RJ, Ausiello DA, Cantiello HF. The cystic fibrosis transmembrane conductance regulator is a dual ATP and chloride channel. J Biol Chem 1994; 269: 20584-20591.

8 Hazama A, Miwa A, Shimizu T, Yamamoto T, Okada Y. Osmotic swelling-induced ATP release from cultured intestinal epithelial cells. Jpn J Physiol 1998; 48: Suppl. (in press)

9 Kubo M, Okada Y. Volume-regulatory Cl^- channel current in cultured human epithelial cells. J Physiol 1992; 456: 351-371.

10 Yada T, Oiki S, Ueda S, Okada Y. Intestinal secretagogues increase cytosolic free Ca^{2+} concentration and K^+ conductance in a human intestinal epithelial cell line. J Membr Biol 1989; 112: 159-167.

11 Tsumura T, Oiki S, Ueda S, Okuma M, Okada Y. Sensitivity of volume-sensitive Cl^- conductance in human epithelial cells to extracellular nucleotides. Am J Physiol 1996; 271: C1872-C1878.

12 Schwiebert EM, Egan ME, Hwang TH, Fulmer SB, Allen SS, Cutting GR, Guggino WB. CFTR regulates outwardly rectifying chloride channels through an autocrine mechanism involving ATP. Cell 1995; 81: 1063-1073.

13 Hazama A, Miyoshi T, Hayashi S, Miwa A, Yamamoto T, Okada Y. CFTR Cl^- channel and epithelial Cl^- and ATP secretion. Jpn J Physiol 1997; 47: Suppl. 2, S24.

14 Liu Y, Tsumura T, Oiki S, Okada Y. Glibenclamide inhibits the volume-sensitive Cl^- conductance in human epithelial cells. Jpn J Physiol 1997; 47: Suppl. 2, S123.

15 Strong TV, Boehm K, Collins FS. Localization of cystic fibrosis transmembrane conductance regulator mRNA in the human gastrointestinal tract by in situ hybridization. J Clin Invest 1994; 93: 347-354.

Cell Volume Regulation: The Molecular Mechanism and Volume Sensing Machinery
Y. Okada, editor.

Molecular mechanism of stress-induced cardiac hypertrophy

Issei Komuro

Molecular Cardiology Division, Department of Medicine III,
University of Tokyo School of Medicine, Tokyo 113, Japan

SUMARRY

Mechanical stress is a major cause for cardiac hypertrophy. Although the mechanisms by which mechanical load induces cardiomyocyte hypertrophy have long been a subject of great interest for cardiologists, the lack of a good *in vitro* system has hampered the understanding of the biochemical mechanisms. For these past several years, however, an *in vitro* neonatal cardiocyte culture system has made it possible to examine the biochemical basis for the signal transduction of mechanical stress. Passive stretch of cardiac myocytes cultured on silicone membranes activates phosphorylation cascades of many protein kinases including protein kinase C, Raf-1 kinase and extracellular signal regulated kinases, and induces the expression of specific genes as well as an increase in protein synthesis. During that process, secretion and production of vasoactive peptides such as angiotensin II and endothelin, are increased and they play critical roles in the induction of these hypertrophic responses. However, we have recently obtained evidence suggesting that the vasoactive peptides are not indispensable for the development of mechanical stress-induced hypertrophic responses. The most important question regarding how mechanical stimulus is converted into biochemical signals remains unknown.

INTRODUCTION

Mechanical load is a major factor for cardiac hypertrophy at the pressure or volume overload (1, references therein). Recently there has been increasing evidence to suggest that mechanical stress itself evokes a variety of intracellular biochemical signals and induces the expression of specific genes as well as increasing rates of protein synthesis in cardiac myocytes (for a review ref. 2, 3). In the present short review, we focus on the role of vasoactive peptides in the development of cardiac hypertrophy.

MECHANICAL STRESS IS A PRIMARY STIMULUS FOR CARDIAC HYPERTROPHY INDUCED BY HEMODYNAMIC OVERLOAD

It has been controversial whether a primary stimulus for cardiac hypertrophy is a mechanical stress itself or an accompanying increase in neural or humoral factors. Many studies have shown that the adrenoceptor activation accompanies cardiac hypertrophy induced by hemodynamic overload (4). However evidence presented below suggests that the direct mechanical stress might be a primary factor for cardiac hypertrophy in response to hemodynamic overload.

First, cardiac hypertrophy can be induced by hemodynamic overload even after adrenoceptor blockade or sympathectomy (1). Second, increased cardiac load stimulates protein synthesis in isolated hearts (5). Third, it has been shown that stretching the cultured cardiomyocytes with no involvement of neural or humoral factors stimulates protein synthesis and specific gene expression (6-8).

MECHANICAL STRESS CAN BE APPLIED TO CARDIOMYOCYTES BY STRETCHING THE CELL

How is mechanical stress percieved by a cell as a stimulus? With the use of a Langendorf preparation, it has been shown that stretch of the ventricular wall as a consequence of increased aortic pressure is the mechanical parameter most closely related to the increase in protein synthesis (5). This observation has been confirmed by the experiments of cardiocytes cultured *in vitro* with serum-free media. When cardiocytes cultured on deformable silicone membrane were stretched, an increase in protein synthesis was observed (6). Stretching the cardiocytes also induced the expression of specific genes, such as immediate early genes (IEG) and fetal type genes (for a review see ref. 2, 3). These gene expression patterns are similar to those observed in the heart *in vivo* in response to hemodynamic overload. These observations suggest that mechanical stress (hemodynamic overload) might have effects on cardiocytes as the result of stretch.

MECHANICAL STRESS ACTIVATES A PROTEIN KINASE CASCADE OF PHOSPHORYLATION

We first found that mechanical stretch of cultured cardiac myocytes strongly induced expression of IEG such as c-fos and c-jun genes (7). Since expression of c-fos gene was so strong, we analyzed the element of c-fos gene promoter which responds to mechanical stress to get insights into the responsible signals. Serum response factor binding element (SRE) of the c-fos gene promoter is important for the response to mechanical stress, suggesting that protein kinase C (PKC) may be involved in the stretch-induced c-fos gene expression. Actually, PKC was activated by stretch in cardiac myocytes (8).

A great number of intracellular signals are transduced into a nucleus through a protein kinase cascade of phosphorylation (9). Once a mechanical stimulus is converted into biochemical signals involving phosphorylation events, how do these signals then regulate protein synthesis and gene expression in cardiac myocytes? A growing body of evidence has suggested that extracellular signal regulated kinases (ERKs) play a critical role in cell proliferation and differentiation (10). ERKs have been reported to be important for the development of hypertrophic responses including specific gene expression and an increase in protein synthesis (11, 12). Myocyte stretching activated ERKs and its upstream kinase Raf-1 kinase (Fig. 1)(13-15). Activation of PKC was known to increase the activity of Raf-1 kinase and ERKs in various cell types (10). When PKC was depleted by preincubating myocytes with phorbol ester or its activity was blocked by the PKC-inhibitors, stretch-induced ERK activity was decreased by more than 80%, suggesting that the stretch-induced increase in ERK activity occurs mainly through PKC-dependent pathway. The stretch-induced ERK activation is not dependent on Src family or Ras.

Figure 1. Phosphorylation Cascades Activated by Mechanical Stress
Mechanical stress induces many events including gene expressions and hypertrophy in the cardiac myocytes through protein kinase cascade of phosphorylation. PKC, protein kinase C; MEK, MAP kinase kinase; ERKs, extracellular signal regulated kinases.

Mechanical stress → ? → PKC → Raf-1 → MEK → ERKs → protein synthesis / gene expression

AUTOCRINE AND PARACRINE MECHANISMS

Second messenger cascades activated by mechanical stress on cardiac myocytes are very similar to those evoked by the addition of growth factors and cytokines. Cardiac myocytes and non-muscle cells such as fibroblasts, endothelial cells and smooth muscle cells may secrete some hypertrophy-promoting factors following a stretch stimilus. Mechanical stress on myocardial cells has been reported to increase the synthesis of some growth-promoting factors (16). Many growth factors, including acidic and basic FGF, TGFβ, insulin-like growth factor I and II, have been reported to exist in the heart, and some of them can induce cardiac hypertrophy and specific gene expression in the cultured cardiomyocytes (for a review see ref. 17). However, there have been only a few reports regarding the expression of these growth factors during hypertrophy induced by hemodynamic overload.

A growing body of evidence has suggested that the local renin-angiotensin system is important for cardiac hypertrophy (for a review see ref. 18). All components of the renin-angiotensin system such as angiotensinogen, renin and angiotensin converting enzyme (ACE) have been identified at both the mRNA and protein levels in the heart. Angiotensin II (AII) stimulates protein synthesis in cultured cardiomyocytes. Increases in angiotensinogen and ACE mRNAs have been reported in hypertrophied left ventricle of rats. Subpressor doses of ACE inhibitors can prevent or cause regression of cardiac hypertrophy with no change in systemic systolic blood pressure. An increase in left ventricular mass which was produced by abdominal aortic constriction, without significant increase in plasma renin activity, was completely prevented with an ACE inhibitor without any change in afterload (19). These results suggest that the local renin-angiotensin system may play a critical role in cardiac hypertrophy induced by pressure overload.

AII increases c-*fos* gene expression as well as protein synthesis in neonatal rat cardiocytes through AII type 1 receptor (AT1). Up-regulation of c-*fos* gene expression by AII is blocked by PKC inhibitors, and AII actually increases production of inositol phosphates and activates PKC (20, 21). These signals elicited by AII in cardiac myocytes are very similar to those evoked by mechanical stress as mentioned above. Therefore, we examined the possible involvement of AII in the stretch-induced increase in protein synthesis. Stretching cardiocytes induced a 1.5-fold increase in protein synthesis as described previously (7). The addition of AT1 antagonists partially blocked the stretch-induced increase in amino acid incorporation. The induction of c-*fos* gene expression and the activation of ERKs by stretching were also attenuated by the antagonist (14, 22, 23). It was reported that mechanical stretch causes a release of AII from cultured cardiocytes (22). These results suggest that AII is responsible in part for events evoked by mechanical stress on cardiac myocytes.

Endothelin-1 (ET) has been reported to induce cardiomyocyte hypertrophy with enhanced expression of fetal genes. We next examined the involvement of ET in the development of stretch-induced cardiomyocyte hypertrophy. The pretreatment with BQ123, a specific ET typeA receptor antagonist, significantly suppressed the induction of hypertrophic responses by stretch (24). These results suggest that vasoactive peptides such as AII and ET are critically involved in the development of mechanical stress-induced cardiac hypertrophy.

VASOACTIVE PEPTIDES ARE NOT INDISPENSABLE FOR THE DEVELOPMENT OF CARDIAC HYPERTROPHY

Stress-Activted Protein Kinase (SAPK)

Several members of MAP kinase family have been isolated and among them, the stress-activated protein kinases (SAPKs)/c-Jun NH_2-terminal kinases as well as ERKs have been characterized well. SAPKs are the mammalian homologs of yeast Hog-1, which is activated by osmotic stress. SAPKs are activated by a variety of cellular stresses such as heat shock, ultra violet irradiation and ischemia/reperfusion. When cardiac myocytes of neonatal rats cultured on a deformable silicone dish were stretched, the activity of SAPKs was increased from 10 min, peaked at 30 min and gradually decreased thereafter. The increase in the activity of SAPKs was proportional to the stretch. Unlike ERKs, the activation of SAPKs by stretching cardiac myocytes was not suppressed by the pretreatment with AT1 antagonists, suggesting that there are some signals which are independent of the secreted AII (25).

AII Knockout Mice

Using AT1a knockout (KO) mice, where signals through AT1 are genetically deleted, we examined whether AII is indispensable for the development of pressure overload-induced acute hypertrophic responses. Reverse-transcriptase polymerase chain reaction analysis revealed that AT1 mRNA levels were less than 10% in the heart of KO mice compared with wild type (WT) mice, while AT2 gene was expressed at almost same levels in the heart of both mice. Acute pressure overload produced by constricting transverse aorta induced expressions of IEG and activations of ERKs in the heart of KO mice as well as WT mice. Both basal and activated levels of all these responses were significantly higher in KO mice than WT mice. Chronic pressure overload for 30 days produced by constriction of the abdominal aorta markedly increased the heart weight to body weight ratio and induced fetal gene expressions in both mice at almost the same degree. These results suggest that cardiac hypertrophy could be induced by pressure overload in *in vivo* heart without AT1a signaling.

We next examined the hypertrophic responses using cultured cardiac myocytes prepared from AT1a KO mice to elucidate the molecular mechanism of the development of hypertrophic responses without AT1 signals. When cardiac myocytes were stretched by 20% for 10 min, ERKs were strongly activated in KO cardiomyocytes as well as WT myocytes. Both basal and stimulated levels of ERKs were higher in cardiomyocytes of KO mice than of WT mice. An AT1 antagonist attenuated stretch-induced ERK activation in WT myocytes but not in KO myocytes. Downregulation of protein kinase C inhibited stretch-induced ERK activation in WT cardiocytes, while a tyrosine kinase inhibitor, genistein, suppressed the ERK activation in KO cardiac myocytes. These results suggest that mechanical stretch could evoke hypertrophic responses in cardiac myocytes which lack AT1 signaling possibly through tyrosine kinase activation.

MECHANISMS BY WHICH MECHANICAL STRESS IS CONVERTED INTO A BIOCHEMICAL STIMULUS

Stretch-Sensitive Ion Channels

Many cells respond to a variety of environmental stimuli by ion channels in the plasma membrane. Mechanosensitive ion channels have been observed with single-channel recordings in more than 30 cell types of prokaryotes, plants, fungi and all animals so far examined (for a review see ref. 26). The activation of stretch-sensitive channels has been proposed as the transduction mechanism between load and protein synthesis in cardiac hypertrophy (27). The stretch-sensitive channels allow the passage of the major monovalent physiological cations, Na^+ and K^+, and the divalent cation, Ca^{2+}. With the use of a Ca^{2+}-binding fluorescent dye (fluo3) and the patch clamp technique, mechanically induced Ca^{2+} influx through stretch channels was shown to lead to waves of calcium-induced calcium release (28).

Inhibitors of stretch-sensitive cation channnels (gadolinium and streptomycin), of ATP-sensitive K^+ channels (glibenclamide), of hyperpolarization-activated inward channels (CsCl) did not inhibit IEG expression and protein synthesis by stretching (unpublished results). Preincubation with Ca^{2+} channel blockers or short exposure to EDTA also did not change the induction of c-*fos* mRNA after stretching. The induction of c-*fos* expression was observed even in Na^+-free (Na^+ replaced by choline) or Ca^{2+}-free media. Ion influx through stretch sensitive channels may play certain roles such as stretch-induced arrhythmias (29). Although we cannot rule out the existence of the inhibitor-insensitive stretch channels in cardiomyocytes, the many biochemical events that are evoked by stretching cardiac myocytes cannot be explained by an opening of the stretch-sensitive ion channels.

Extracellular Matrix and Cytoskeleton

Many data have been accumulated indicating that the mechanical stress is transduced into the cell from the sites at which cells attach to extracellular matrix (ECM)(for a review see ref. 30). Transmembrane ECM receptors, such as the integrin family, are good candidates for mechanoreceptors. A large extracellular domain of integrin receptor complex binds various ECM proteins, while a short cytoplasmic domain has been shown to interact with the cytoskeleton in the cell (for a review see ref. 31). Since cytoskeleton proteins can potentially regulate plasma membrane proteins such as enzymes, ion channels and antiporters, mechanical stress could modulate these membrane associated proteins and stimulate second messenger systems through the cytoskeleton. Integrins can transmit signals not only by organizing the cytoskeleton but also by altering biochemical properties such as the extent of tyrosine phosphorylation of a complex of proteins including pp125^{fak} (30). Rho small G proteins are important for the function of integrins through regulating actin cytoskeleton. When functions of Rho family was inhibited by C3 exoenzyme or overexpression of Rho GDP dissociation inhibitor and dominant negative constructs, stretch-induced activation of ERKs and expression of IEG were strongly suppressed (unpublished results). In contrast, AII-induced hypertrophic responses were not suppressed by these treatments, suggesting that Rho family and possibly integrin play a critical role in mechanical stress-induced hypertrophic responses. How Rho or integrin is involved in the stretch-induced hypertrophic responses remains to be determined.

FUTURE DIRECTIONS

As shown in Figure 1, mechanical stress can evoke a variety of signals in cardiac myocytes, and the molecules which are involved in the signal transduction pathway of mechanical stress are similar to the molecules which play important roles in many other cells stimulated by growth factors. Recently, yeast genes encoding members of MAP kinase have been isolated by complementation of yeast mutations as an essential protein for restoring the osmotic gradient across the cell membrane in response to increased external osmorality (32). This suggests that once cells receive external stimuli, intracellular signal transduction pathways are usually highly conserved among many cell types and many species.

Although many biochemical events which occur in cardiac myocytes subsequent to mechanical stretch have been clarified, a main intriguing question remains unanswered. How is mechanical stress converted into biochemical signals? In other words, what is the mechanoreceptor or the transducer for mechanical stress in cardiac myocytes? It is conceivable that by stretching the plasma membrane, mechanical stress directly changes the conformations of the functional proteins such as enzymes and G-proteins, or directly activates enzymes such as phospholipase by physically placing the enzymes close to their phospholipid substances in the plasma membrane. As mentioned above, the integrin-cytoskeleton complex seems to be an alternate candidate structure for a mechanoreceptor and a transducer. Integrin-cytoskeleton proteins not only play a "passive" role such as maintaining the cell structure, but also may play "dynamic" roles in regulation of cellular functions such as protein phosphorylation and activation of an antiporter. Integrin $\alpha_5\beta_1$ was shown to activate the Na^+/H^+ antiporter by binding to fibronectin, suggesting that integrin can behave similarly to a growth factor receptor in activating signaling pathways (33). The cytoskeleton has been also shown to play an important role in secretion. Mechanical stress may stimulate secretion of some cytokines, and these cytokines may generate multiple intracellular signals as a secondary event. Further studies to identify specific signaling molecules, including mechanoreceptors and mechanotransducers, and characterization of their activities will be required for understanding of the physiological functions of mechanical stimuli on cardiac myocytes and to finally clarify the mechanisms by which adaptive cardiac hypertrophy deteriorates into congestive heart failure.

REFERENCES

1 Cooper G, Kent RL, Uboh CE, Thompson EW, Marino TA. J Clin Invest 1985;75:1403-1414.

2 Komuro I, Yazaki Y. Annu Rev Physiol 1993;55:55-75.

3 Komuro I, Yazaki Y. Trends in Cardiovascular Medicine 1994;4:117-121.

4 Zak R. Zak R, ed. New York, Raven Press. 1984;pp165-185.

5 Kira Y, Kochel PJ, Gordon EE, Morgan HE. Am J Physiol 1984;246:C247-C258.

6 Mann DL, Kent RL, Cooper G. Circ Res 1989;64:1079-1090.

7 Komuro I, Kaida T, Shibasaki Y, Kurabayashi M, Katoh Y, Hoh E, Takaku F, Yazaki Y. J Biol Chem 1990;265:3595-3598.

8 Komuro I, Katoh Y, Kaida T, Shibasaki Y, Kurabayashi M, Takaku F, Yazaki Y. J Biol Chem 1991;266:1265-1268.

9 Cantley LC, Auger KR, Carpenter C, Duckworth B, Graziani A, Kapeller R, Soltoff S. Cell 1991;64:281-302.

10 Blenis J. Cancer Cells 1991;3:445-449.

11 Thorburn J, Frost JA, Thorburn J. J Cell Biol 1994;126:1565-1572.

12 Glennon PE, Kaddoura S, Sale EM, Sale GJ, Fuller SJ, Sugden PH. Circ Res 1996;78:954-961.

13 Yamazaki T, Tobe K, Hoh E, Maemura K, Kaida T, Komuro I, Tamemoto H, Kadowaki T, Nagai R, Yazaki Y. J Biol Chem 1993;268:12069-12076.

14 Yamazaki T, Komuro I, Shiojima I, Mizuno T, Kudoh S, Zou Y, Takano H, Hiroi Y, Ueki K, Tobe K, Kadowaki T, Nagai R, Yazaki Y. J Clin Invest 1995;96:438-446.

15 Sadoshima J, Izumo S. EMBO J 1993;12:1681-1692.

16 Hammond GL, Wieben E, Market CL. Proc Natl Acad Sci USA 1979;76:2455-2459.

17 Parker TG, Schneider MD. Annu Rev Physiol 1991;53:179-200.

18 Baker KM, Booz GW, Dostal DE. Annu Rev Physiol 1992;54:227-241.

19 Baker KM, Cherin MI, Wixon SK, Aceto JF. Am J Physiol 1990;259:H324-H332.

20 Katoh Y, Komuro I, Shibasaki Y, Yamaguchi H, Yazaki Y. [abst] Circulation 1989;80:II-450

21 Sadoshima J, Izumo S. Circ Res 1993;73:424-438.

22 Sadoshima J, Xu Y, Slayter HS, Izumo S. Cell 1993;95:977-984.

23 Yamazaki T, Komuro I, Kudoh S, Zou Y, Shiojima I, Mizuno T, Takano H, Hiroi Y, Ueki K, Tobe K, Kadowaki T, Nagai R, Yazaki Y. Circ Res 1995;77:258-265.

24 Yamazaki T, Komuro I, Kudoh S, Zou Y, Shiojima I, Hiroi Y, Mizuno T, Maemura K, Kurihara H, Aikawa R, Takano H, Yazaki Y. J Biol Chem 1996;271:3221-3228.

25 Komuro I, Kudoh S, Yamazaki T, Zou Y, Shiojima I, Yazaki Y. FASEB J 1996;10:631-636.

26 Morris CE. J Memb Biol 1990;113:93-107.

27 Bustamante JO, Ruknudin A, Sachs F. J Cardiovasc Pharmacol 1991;17(Suppl. 2):S110-S113.

28 Sigurdson W, Ruknudin A, Sachs F. Am J Physiol 1992;262:H1110-H1115.

29 Hansen DE, Borganelli M, Stacy GP, Taylor LK. Circ Res 1991;69:820-831.

30 Juliano RL, Haskill S. J Cell Biol 1993;120:577-585.

31 Hynes RO. Cell 1992;69:11-25.

32 Brewster JL, de Valoir T, Dwyer ND, Winter E, Gustin MC. Science 1993;259:1760-1763.

33 Schwartz MA, Lechene C, Ingber DE. Proc Natl Acad Sci USA 1991;88: 7849-7853.

Part III
Ion Channel and Transporter Involved in Volume Regulation

Cell Volume Regulation: The Molecular Mechanism and Volume Sensing Machinery
Y. Okada, editor.

Cell volume regulation: the role of chloride channels

Tim J.C. Jacob, Liwei Wang and Lixin Chen

School of Molecular and Medical Biosciences, University of Wales, Cardiff CF1 3US, UK.

VOLUME CONTROL AND OSMOLYTES

All cells regulate their volume. Control of volume is essential to survival and a loss of control results in cell death. The mechanisms involved in cell volume regulation developed very early in evolution and, ever economical, nature has borrowed mechanisms that contribute to other cellular functions. To drive cell division and provide turgor pressure in bacteria and plants, cells must accumulate solutes. The transport proteins that are responsible for uptake of solutes belong to very old gene families indeed and must have been among the first membrane proteins to evolve, providing a selective entry route for essential metabolites. Some of those transporters now have a role in volume regulation. The solutes they transport are compatible with biological processes. Cations are also utilised in volume regulation but high concentrations usually perturb cellular function so, although they are used as short-term osmolytes, other, non-polar solutes are usually preferred. Throughout the phylogenetic kingdoms only a few types of osmolytes are utilised as compatible solutes:

- quarternary amines (e.g. glycine betaine)
- amino acids (e.g. taurine, alanine, glutamate, proline)
- polyols (e.g. sorbitol, myo-inositol).

In the process of responding to increases in cell volume, cells efflux these osmolytes through specific membrane protein channels. In lower eukaryotes this seems to be the predominant mechanism for volume adjustments but, in higher eukaryotes, anions, in particular Cl^-, are also utilized. The question currently being addressed in the study of volume regulation is whether the same, or different, channels are involved in both processes.

VOLUME "SET POINT"

Research into the processes of volume regulation has tended to consider volume increase and volume decrease as separate and distinct mechanisms. Large step increases in osmolarity cause the cell to shrink, activating regulatory volume increase (RVI) and large step decreases in osmolarity cause the cell to swell, activating regulatory volume decrease (RVD). However, in reality cell volume changes will mostly be small and gradual. There is evidence that, when osmolarity is changed gradually, the volume regulatory mechanisms can keep pace with the change and cell volume does not change. When this occurs different mechanisms for maintaining volume are activated than those activated during RVD (1). Regulatory volume increase involves the process of loading osmolytes (mentioned above) and this is probably going on all the time. It will be modulated by changes in state - for example increases and decreases in cell volume. Conversely, to counteract the natural tendency of cells to swell (due to oncotic pressure, catabolism and the entry of ions/metabolites via cotransporters), material will need to be lost from the cell. The RVD mechansism(s) will therefore need to be up- and down-regulated to meet the needs of the cell at any moment in time. Cells appear to have "set points" at which various mechanisms are switched on or off. For example, the activation of the volume sensitive organic osmolyte and anion channel (VSOAC) is regulated by the level of intracellular electrolytes (2) and, in rat hepatocytes, cAMP appears to activate a swelling-activated Cl^- conductance by altering the volume set point (3). The metabolic state of the cell will also influence the way in which the cell regulates its volume since activation of the volume regulated anion channel(s) is dependent upon nonhydrolytic ATP binding (4). Accumulation of organic osmolytes is metabolically expensive and thus it may be advantageous to the cell to lose Cl^- if levels are sufficiently high (5). The swelling-induced loss of organic osmolytes is inhibited by elevated intracellular ionic strength and, in particular, by elevated intracellular Cl^- (6, 7). This may have the advantage that, when intracellular Cl^- levels are high, the loss of expensive organic osmolytes is inhibited and, instead, the osmolarity is restored by loss of Cl^-, either through the same, organic osmolyte/anion channel or through separate anion-selective channels. A counterion, usually K^+, will accompany this efflux of Cl^-. How the cell responds to volume changes will also depend upon the nature of the volume stress, for example the response to prolonged osmotic stress may well be accompanied by gene transcription and increased levels of mRNA for proteins actively involved in volume regulation and its control.

OSMOTIC REGULATION OF GENE TRANSCRIPTION

In response to hyperosmotic stress, bacteria initially lose water and their volume decreases. There is a loss of turgor. To restore turgor the cell increases the concentration of cytoplasmic solutes, primarily by accumulating K^+ and synthesising glutamate to act as a charge neutralising counterion. Potassium glutamate may then trigger the transcription of a number of genes (see Fig. 1).

STIMULUS[1] → SIGNAL[2] → EFFECTOR PROTEIN[3] → ACTIVATOR[4] → PHYSIOLOGICAL EFFECT[5]

Figure 1. Osmotic regulation of gene transcription

[1]Stimulus = osmotic shock
[2]Unknown, but see below
[3]Effector protein, its DNA binding is specificity is altered, possibly by phosphorylation. RNA pol ability to express target gene altered
[4]A positive activator protein can enhance transcription from a weak promoter or a repressor can inhibit transcription from a strong promoter
[5]Gene transcription, increased levels of volume regulatory proteins. [Modified from ref. 8]

Two-dimensional gel electrophoresis shows that synthesis of more than 22 proteins is induced by hyperosmotic stress in bacteria. In addition, 11 or 12 genetic loci have been identified as being regulated by elevated osmolarity (8).

Ionic strength contributes to gene regulation in mammalian cells. In rabbit inner medulla cells, sorbitol is accumulated during NaCl stress. When NaCl increases in the growth medium, the levels of aldose reductase transcription increase, resulting in an increase in aldose reductase (AR) - the enzyme which converts glucose to sorbitol (9). AR transcription correlates with cellular ionic strength, particularly with the sum of cellular Na^+ and K^+ (10). When MDCK cells are exposed to hypertonic medium the transcription of genes coding for osmolyte uptake transporters is increased. The levels of Na^+/myo-inositol mRNA increase allowing for increased uptake of myo-inositol (11) and exactly the same happens for the Na^+/Cl^-/betaine transporter (12).

OSMOTIC SWELLING AND RVD

In protozoans, osmotic swelling induces the efflux of anions and organic osmolytes. In *Acanthamoeba castellani*, the response to osmotic change is to change the internal levels of free amino acids, in particular alanine, glycine and proline, with intracellular ions making a relatively small contribution to the cell's adaptation to the altered osmotic environment (13). In response to hypoosmotic stress there was a three-fold greater loss of amino acids than inorganic ions in the short-term (60min), and thereafter, only the amino acid pool decreased any further (13, 14). When the protozoan parasite *Giardia intestinalis* was exposed to hypotonic medium it underwent RVD that involved the rapid release of neutral intracellular amino acids, especially alanine. This volume-activated alanine efflux was sensitive to temperature ($Q_{10} = 3.6$), suggesting carrier-type transport, it was not inhibited by the anion channel blockers (furosemide, iodoacetate, niflumate) but was blocked by the thiol-reagent, para-hydroxymercuribenzoate (PHMB).

In the unicellular protozoan parasite, *Crithidia luciliae*, RVD was accompanied by the efflux of amino acids (predominantly alanine, proline and glycine) (15). The efflux of electroneutral amino acids was greater than that of the anionic amino acid glutamate. The efflux was inhibited by N-ethylmaleimide but not by a range of anion transport blockers (DIDS, NPPB, quinine, tamoxifen, niflumate). The kinetics were those of a channel rather than a transporter but the pharmacology of the system differs from that of the anion-selective channels that are thought to mediate the volume-regulatory efflux of organic osmolytes from vertebrate cells (15).

In mammalian cells the contribution of osmolytes to the process of RVD is variable. In MDCK cells amino acid and Cl^- loss were about equal (16), there was a significant efflux of taurine in rat brain (17) and mammary gland (18), but in other studies it was concluded that the contribution of osmolytes during RVD was relatively small in comparison to inorganic ions (19, 20, 21).

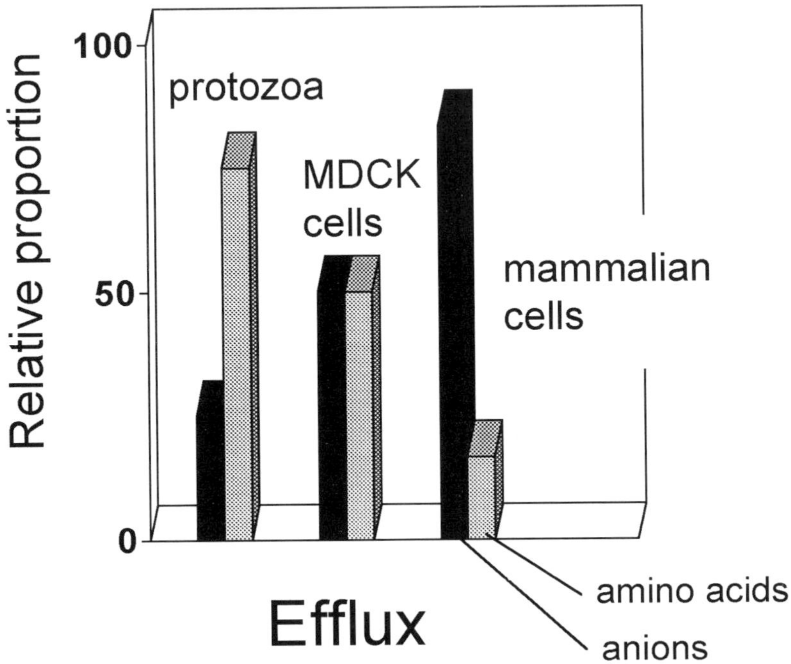

Figure 2. The relative proportion of anion versus organic osmolyte efflux in response to an increase in volume has changed over evolutionary time. In primitive organisms (e.g. protozoans), organic osmolyte efflux is the predominant response to osmotic swelling whereas, in mammalian cells, anion efflux predominates. However, this response changes with time of exposure to a hypotonic challenge and will also depend upon the initial conditions of the cells (see text).

If the same channel family is responsible for mediating organic osmolyte efflux in protozoa, lower eukaryotes and through to higher vertebrates then, in the course of evolution, later members of the channel family must have acquired both a Cl^- permeability and the capacity to regulate this permeability.

ACTIVATION OF SWELLING-ACTIVATED ANION CHANNEL

There have been a number of proposals for the "switch" or "signal" that transduces volume increase into increased anion/osmolyte conductance;

i) membrane stretch triggering a second messenger pathway

- arachidonic acid metabolism (22, 23)

- leukotriene LTD_4 (24)
- G-protein coupling (25)
- kinase activity (22, 26)
- calcium influx through stretch-activated cation channels (27)

ii) direct involvement of cytoskeleton
- actin (28, 29)
- tensegrity (30)
- anchor-insertion of channel protein (5)

iii) reduction in "macromolecular crowding"
- changes in concentration of cytosolic regulatory proteins (31)

NATURE OF CHANNEL/CHANNEL ACTIVATOR(S)

The fact that the mRNA coding for putative volume-activated ion channels from one species can be injected into Xenopus oocytes and, by so doing, a new volume-activated anion channel phenotype can be conferred suggests that, either there is a conservation of the second messenger activation pathway between unrelated species or, that the channel protein itself contains the sensor. In the case of ClC-2, a voltage-gated Cl^- channel cloned from rat brain (32), deletion mutants could be constructed that produced either constitutively active channels or channels that were unresponsive to volume changes (33). "Intermediate" mutants were also produced that were constitutively open and yet were still responsive to volume changes. The constitutive activity of these channels could by further stimulated by cell swelling, and, unlike both the wild-type and "open" mutants, the constitutive activity decreased with cell shrinkage. The results suggest the channel can be decoupled from any second messenger pathway, remaining fixed in either the open or closed state. The activity of the intermediate mutant could be interpreted as an altered sensitivity to the endogenous second messenger or, less likely, as a decoupled, but self-regulating state where the channel itself is able to "sense" volume changes. In view of the unlikely nature of the latter interpretation one is left with the conclusion that a second messenger pathway exists in Xenopus oocytes that activated exogenous ClC-2.

In spite of the fact that ClC-2 has been demonstrated to be a Cl^- channel activated by cell swelling, it exhibits properties that are unlike those of the volume-activated Cl^- channels expressed in most cells so far examined. For this reason it is not thought to be the volume-activated anion channel (34).

Recent evidence suggests that another member of the ClC family, ClC-3, may be the channel responsible for the volume-activated chloride current, $I_{Cl,vol}$ (35). Expression of a cardiac clone of ClC-3 in NIH/3T3 cells resulted in a large basally active chloride conductance that was modulated by cell volume and expressed many properties of $I_{Cl,vol}$ observed in native cells. Furthermore a mutation at position 579 abolished the outward rectification and altered the anion selectivity from $I^- > Cl^-$ to $Cl^- > I^-$ while leaving the swelling activation intact (35).

Using mRNA from Madin Darby canine kidney (MDCK) cells expressed in Xenopus oocytes Paulmichl et al. (36) identified a new Cl^- channel which they called I_{Cln}. It generated a Cl^- channel displaying the features of swelling-induced Cl^- currents in many other cells (including those of both epithelial and muscle tissues) but the I_{Cln} current was active in the absence of any osmotic challenge and the primary structure of the protein itself did not suggest any conventional transmembrane domains. Subsequent studies have demonstrated a predominantly cytosolic localisation of the protein and have suggested that the expression of I_{Cln} in Xenopus oocytes activated the endogenous swelling-activated Cl^- channel (37). Using monoclonal antibodies against I_{Cln}, injected into oocytes, it was shown that the endogenous swelling-activated Cl^- current in oocytes, identified by Ackerman et al., (38) and termed $I_{Cl,swell}$, was abolished (37). They suggested that the protein-protein interaction between I_{Cln} and soluble actin and the recognition of cytoskeletal-associated I_{Cln} makes I_{Cln} a candidate for a regulatory scheme involving the cytoskeleton. Voets et al. (39) have concluded that I_{Cln} currents were distinct from $I_{Cl,\,swell}$ in Xenopus oocytes on the grounds of differences in nucleotide block, rectification and ionic selectivity. Strange et al (5) have put forward the case for the identity of I_{Cln} as VSOAC circumventing the problem of its cytoplasmic location by proposing an "anchor-insertion" model (40) for which there is some support (41) but also some negative evidence (42). Szucs et al. (43) have suggested that "pI_{Cln} is a good molecular candidate for $I_{Cl,vol}$ in endothelial cells" on the basis of kinetic properties, ionic selectivity, and block by external ATP, AZT, gossypol and acyclovir. Paulmichl and co-workers have recently reconstituted purified I_{Cln} into lipid bilayers and observed channel activity (Paulmichl, personal communication) strengthening the case for its identity as one of the volume-activated anion channels.

Another candidate protein for the role of swelling-activated Cl^- channel is P-glycoprotein (P-gp) (4, 44) although this has often been challenged (see below). In this context is is interesting to note that unlike I_{Cln} and ClC-2, the other strong contenders for this role, P-gp does not produce swelling-activated, or any other, Cl^- channel activity when injected into Xenopus oocytes (45). This particular study was, however, carried out on oocytes without an endogenous volume-activated Cl^- current and the authors point out that Xenopus oocytes may be without a necessary regulatory protein and it is not known whether human P-gp can interact with the Xenopus signalling system.

P-GLYCOPROTEIN: CHANNEL OR REGULATOR?

P-gp is a hydrophobic drug pump that confers the multidrug resistance phenotype when overexpressed (46). Whether there is a relationship between P-gp and volume-activated anion channels has been the subject of much debate. Initially it was proposed that P-gp was the channel itself (4, 47) but subsequent studies have produced conflicting evidence for the relationship between the expression of P-gp and volume-activated Cl^- currents. An analysis of the studies that have investigated this relationship demonstrates that the choice of biological system often dictates the outcome (Table 1).

Table 1. The relationship between P-glycoprotein and volume-activated Cl^- currents.

P-gp is associated with $I_{Cl,swell}$	Cell system	Expression system	Ref.
√	NIH3T3 fibroblasts	transient transfection with MDR1 cDNA, colchicin selection	4, 47
√*	human breast cancer cells	MDR1 transfection, vinblastine selection	48
√	native bovine epithelial cells	endogenous P-gp	49, 50, 51 , 52
√	NIH3T3	colchicin selection	53
√*	(i) Chinese hamster lung fibroblast, LZ-8 (ii)BALB/c-3T3	(i) doxorubicin selection, (ii) transfection without drug selection	54
√[+]	CHO and NIH3T3 cells	MDR transfection with vinblastine and colchicine selection respectively	47
√	Epidermoid cell line (KB-3-1)	vinblastine	55 1997
-	NIH/3T3	non P-gp expressing[@]	56
-[#]	T84	endogenous P-gp	57
-[$]	Intestine 407 cells & human HepG2	endogenous P-gp	58
x	Human uterine sarcoma	doxorubicin	59
x	(i) Human melanoma FEM-X, (ii) murine lymphoma, (iii) NIH3T3	colchicine	60
x	human leukemia	vinblastine	61
x	rat adenocarcinoma	colchicin	62
x	LZ-8	doxorubicin	63
x	(i) Human colon adenocarcinoma (ii) Human tracheal epithelium	doxorubicin	64
x	NIH3T3	colchicin	65
-	Xenopus oocytes	endogenous	38

√ = positive relationship between P-gp expression or activity and $I_{Cl,swell}$

x = negative relationship between P-gp and $I_{Cl,swell}$ in studies using transformed (cancer) cells or cells induced to overexpress MDR1 by drug selection

- = negative studies conducted in untransformed, cultured cells or Xenopus oocytes.

*No evidence of P-gp expression or of P-gp associated $I_{Cl,swell}$ in untransfected, or parental cells.

[@] NIH3T3 cells used as an example of a cell not expressing P-gp on the basis of drug sensitivity and absence of C219 immunostaining.

[$] C219 failed to inhibit $I_{Cl,swell}$ and antisense ablation of P-gp mRNA had no effect on $I_{Cl,swell}$.

[#] relationship dismissed on basis that P-gp inhibitor (dideoxy-forskolin) failed to inhibit swelling-induced short circuit current

[+] P-gp modulates cell regulatory volume decrease.

In almost all *MDR1* overexpression studies and studies using drug resistant, *MDR1*-expressing cells the authors have failed to find a relationship between P-gp expression and volume-activated anion conductance. One explanation for this is that the levels of P-gp far exceed those required to achieve maximal activation of the volume-activated anion channel(s). Another reason for the contradictory results is that there is more than one volume-activated anion channel and not all the channels are regulated by P-gp. It has been demonstrated that the volume-activated Cl^- current in non-pigmented ciliary epithelial cells is blocked by tamoxifen whereas that of the pigmented cells is not (49, 66) and these cells express multiple, and different, volume-activated Cl^- channels (67). The use of native cells, expressing relatively low levels of P-gp, demonstrates unequivocally, that P-gp is involved in regulating a volume-activated anion channel (49, 52). Firstly, the volume-activated anion channel is inhibited by an antibody to P-gp (49, 48) and secondly, antisense to P-gp mRNA knocks down the volume-activated anion conductance (52).

It is also worth mentioning that, to date, no experiment has ruled out the possibility that P-gp is both a channel and channel regulator. Other examples of this phenomenon exist. For example, CFTR, the cystic fibrosis transmembrane conductance regulator, as well as being a cAMP-dependent Cl^- channel (68) also effects the sensitivity of a renal K^+ channel (ROMK2) to the inhibitory sulfonylurea compound; glibenclamide (69), activates a K^+ conductance when expressed in Xenopus oocytes (70) and downregulates K^+ channel activity when co-expressed with the inward rectifier K^+ channel, KIR6.2 (71).

Recent reviews by Okada (72) and Kirk (73) have raised some specific doubts about the role of P-gp in the regulation of the volume-activated anion current. Okada concludes that P-gp expression is not a prerequisiste to volume-activated anion channel activity (72). The role of "regulator" may not be a prerequisite to channel activity, but, given that there is more than one such channel this conclusion may need to be modified; P-gp may be a prerequisite to one (or more) of the volume-activated anion channels - perhaps playing the same role as the sulphonylurea receptor does to the K_{ATP} ($K_{IR}6.2$) channel (74, 75). Kirk (73) quotes some of the studies in Table 1 as evidence of a lack of correlation between P-gp expression and volume-activated anion channel activity but overlooks much of the supportive evidence and, in addition, refers to the fact that inhibitors of swelling-activated anion channels may not act via P-gp. This has no bearing necessarily on the relationship that does exists between P-gp and volume-activated anion channels and in the light of the foregoing evidence may be explained by the existence of multiple Cl^- current pathways.

We have performed antisense oligonucleotide studies and demonstrated a dose-response relationship between antisense to P-gp mRNA and volume-activated current reduction as well as P-gp expression (52). The antibody inhibition experiments using C219 have been repeated in a number of different cell types, both native and transformed

(48, 49, 54, 76) and only Tominaga et al. (58), in intestinal 407 cells, failed to observe an effect. These latter authors were also unable to demonstrate a correlation between P-gp mRNA knock-down and a reduction in volume-sensitive Cl^- current in intestine 407 cells. More recently the same laboratory has demonstrated that P-gp does indeed modulate a volume-sensitive Cl^- current in a human epidermoid cell line (55).

OTHER CHANNEL CANDIDATES

The $I_{Cl,vol}$ phenotype has been described in many systems, from prokaryotes to mammalian cells, using whole patch clamping and radio-isotope efflux studies. But the problem facing us is that none of the currently cloned candidates genes can account for all aspects of this phenotype. Is there a single gene or are there a number of related genes? The evidence supports the idea that there is a channel, selective for anions and organic osmolytes, that is widely distributed across the phylogenetic kingdoms, and which may have been among the first such mechanisms to evolve. This multi-selective channel appears to be able to modulate its activation set point as well as its relative selectivity such that the anion:organic osmolyte conductance ratio will be varied to suit the prevailing conditions. The identity of this channel is still open to question; I_{Cln} has been proposed (5, 36) but, as already discussed, there are arguments against it. ClC-3 has been convincingly shown to be a volume-regulated Cl^- channel (35), but its selectivity for organic osmolytes is unknown. Other candidates have been suggested, for example, VDAC (voltage dependent anion channel) (77), phospholemman (78, 79), mat-8 (80) and the *band 3* anion exchanger (81, 82). There are arguments against all of these candidates and while it is still possible that they are may contribute to cell volume homeostasis they cannot alone account for the complete phenotype.

ABC TRANSPORTERS - A FAMILY OF REGULATORS

If a single candidate gene for the $I_{Cl,vol}$ phenotype is proving elusive, there is no shortage of candidates for the role of regulator of this phenotype. A number of the ATP-binding cassette (ABC) family of proteins, of which P-gp is a member, are emerging as membrane channel/pump regulators. As we have seen in this review P-gp regulates the volume-activated anion channel and, of the other members of this family, CFTR regulates K^+ channels (69,70, 71), the sulphonylurea receptor (SUR) regulates the K^+_{ATP} channel (74, 75), Pgh1, encoded by the malarial resistance gene, *pfmdr1*, may alter H^+-ATPase activity (83) and other members perform similar functions in bacteria (84). This group of proteins may therefore offer a new level of refinement in the control of membrane transporters and channels.

MULTIPLE CHANNELS AND MULTIPLE REGULATORS

Multiple activator/modulator proteins may exist in the pathway that links cell swelling with anion channel activation; I_{Cln} (85, 86), P-gp (49 Wu et al., 1996), mat-8 (80), phospholemman (78, 79) and IsK (87) are all contenders for this role. Thus it is emerging that, instead of a single, widely-expressed osmolyte/anion channel, there are multiple single channels underlying the swelling-activated Cl^- current. For example, four different types of channel were identified in the basolateral membranes of A6 cells by on-cell patch clamping (88). Of these, three channels, with linear conductances of 12, 30 and 42pS, were characterised on the basis of their i/V relationships; a fourth, outwardly rectifying channel, had inward and outward conductances of 16 and 57pS respectively (88). Using a similar technique Zhang and Jacob (67) have found three swelling-activated Cl^- channels in bovine ciliary epithelial cells, with conductances of 8, 19 and 105pS. The non-pigmented cells exhibited the 8pS and 19pS channels, whereas the pigmented cell exhibited the 8pS and the 105pS channels. The volume-activated Cl^- current of the non-pigmented cells was sensitive to tamoxifen (49) whereas that of the pigmented cells was not (66). On the basis of its sensitivity to tamoxifen it was possible to separate one of these channels, that with a conductance of 19pS, from the other two and this channel had similarities with VSOAC (67).

These separate Cl^- channels may have organic osmolyte permeability but the use of patch clamp recording to identify them has precluded a study of their selectivity to non-polar solutes. So we are no further in associating the $I_{Cl,vol}$ phenotype with a particular gene.

CONCLUSION

There is still no unchallenged candidate for the role of the volume-activated organic osmolyte/anion channel. Indeed it would be surprising if it transpired that there was a single candidate. During evolution there has been a change in the cell's volume regulatory response to swelling. In primitive organisms the solute efflux channel, which mediates cell shrinkage, is selective for organic osmolytes (in particular amino acids). Ascending the evolutionary heirachy to mammalian cells there is a shift to anion efflux as the primary response to cell volume increase, although some organic osmolyte efflux still occurs and may mediate volume regulation in the medium term. There is a continuing debate as to whether the same channel is responsible for both anion and organic osmolyte efflux. Since primitive organisms possess a swelling-activated channel that is selective for organic osmolytes one could argue that, during evolution, this channel either acquired anion selectivity or, that an entirely new class of anion-selective swelling-activated channels emerged. The recent findings of multiple volume-activated chloride channels would favour the latter hypothesis but does not rule out the possibility that there is, in addition, a multi-selective channel. Most likely there is no single gene responsible for the current referred to as $I_{Cl,vol}$. Instead I would suggest that, on the basis of the evidence, there is either a composite multi-selective, multi-conductance membrane channel complex, composed of a number of different single-channel components or, that there a number of separate, independent volume-activated channels. In the former case, the characteristics of the $I_{Cl,\,vol}$ would depend on the number and type of channels already expressed and forming part of the channel complex, while in the latter case, activation, perhaps via linkage to cytoskeletal elements, would be independent and determined by the nature, duration and type of volume stress.

Future molecular studies designed to trace the presence of the solute efflux channel from primitive eukaryotes to higher vertebrates may yield information on how this gene has been modified during evolution and how it contributes to the process of cell volume regulation in mammalian cells.

ACKNOWLEDGEMENTS

I would like to acknowledge other members of the lab who have contributed to the work reported in this review, in particular; Nicky Walker, Jon Stelling, Jin Jun Zhang, Claire Mitchell and Helen Miley. Financial support from the Medical Research Council, The Royal National Institute for the Blind (RNIB), the Wellcome Trust and the Royal Society is gratefully acknowledged.

REFERENCES

1 Van Driessche W, De Smet P, Li J, Allen S et al. Isovolumetric regulation in a distal nephron cell line (A6). Am J Physiol 1997; 272: C1890-C1898.

2 Emma F, McManus M and Strange K. Intracellular clectrolytes regulate the volume set point of the organic osmolyte/anion channel VSOAC. Am J Physiol 1997; 272: C1766-C1775.

3 Meng X-J. and Weinman SA. cAMP- and swelling-activated chloride conductance in rat hepatocytes. Am J Physiol 1996; 271: C112-C120.

4 Gill DR, Hyde SC, Higgins CF, Valverde MA et al. Separation of drug transport and chloride channel functions of the human multidurg resistance P-glycoprotein. Cell 1992; 71: 23-32.

5 Strange K, Emma F and Jackson PS. Cellular and molecular physiology of volume-sensitive anion channels. Am J Physiol 1996; 270: C771-C730.

6 Motais R, Guizouarn H and Garcia-Romeu F. Red cell volume regulation: the pivotal role of ionic strength in controlling swelling-dependent transport systems. Biochim Biophys Acta 1991;1075: 169-180.

7 Jackson PS, Churchwell K, Ballatori N, Boyer JL and Strange K. Single-channel properties of a volume-sensitive anion conductance. Current activation accurs by abrupt switching of channels to an open state. Am J Physiol 1995; 270: C57-C66.

8 Yim HH and Villarejo MR. Gene expression and osmoregulation in bacteria. In: Strange K ed. In Cellular and Molecular Physiology of Cell Volume Regulation. Boca Raton, USA: CRC Press, 1993; 335-346.

9 Smardo Jr FL, Burg, MB and Garcia-Perez AR. Kidney aldose reductase gene transcription is osmotically regulated. Am J Physiol 1992; 262: C776-C782.

10 Uchida S, Garcia-Perez A, Murphy H and Burg M. Signal for induction of aldose reductase in renal medullary cells by high external NaCl. Am J Physiol 1989; 256: C614-C623.

11 Kwon HM. (1993) Molecular regulation of mammalian osmolyte transporters. In: Strange K, ed. The Cellular and Molecular Physiology of Cell Volume Regulation. Boca Raton, USA: CRC Press, 1993; 383-393.

12 Uchida S, Yamauchi A, Preston AS, Kwon HM and Handler JS. Medium tonicity regulates expression of the Na^+- and Cl^--dependent betaine transporter in Madin-

Darby Canine Kidney cells by increasing transcription of the transporter gene. J Clin Invest 1993; 91: 1604-1607.

13 Geoffrion Y and Larochelle J. The free amino acid contribution to osmotic regulation in *Acanthamoeba castellanii*. Can J Zool 1984; 62: 1954-1959.

14 Larochelle J and Gagnon A. Osmoregulation in *Acanamoeba castellanii*. II. Variations of the concentrations of some intracellular ions. Comp Biochem Physiol 1976; 54: 735-379.

15 Bursell JDH, Kirk J, Hall ST, Gero AM and Kirk K. Volume-regulatory amino acid release from the Protozoan parasite *Crithidia luciliae*. J Membr Biol 1992;154: 131-141.

16 Roy G and Sauve R. Effect of anisosmotic media on volume, ion and amino acid content and membrane potential of kidney cells (MDCK) in culture. J Membr Biol 1987; 100: 83-96.

17 Law RO. Taurine efflux and the regulation of cell volume in incubated slices of rat cerebral cortex. Biochim Biophys Acta 1994; 1221: 21-28.

18 Shennan DB, McNeillie SA and Curran DE. Stimulation of taurine efflux from human placental tissue by a hyposmotic challenge. Exp Physiol 1993; 78: 843-846.

19 Hall AC. Cyclic AMP prevents activation of a swelling-induced chloride-sensitive conductance in chick heart cells. J Physiol 1995; 484: 755-766.

20 Brand HS, Meijer AJ, Gustafson LA, Jorning GGA, et al. Cell-swelling-inducedtaurine release from isolated perfused rat liver. Biochem Cell Biol 1994; 72: 8-11.

21 Mountain I, Dclerq PE, Van Driessche W. Volume regulation in rat brain glial cells; lack of substantial contribution of free amino acids. Am J Physiol 1996; 270: C1319-C1325.

22 Civan MM, Coca-Prados M and Peterson-Yantorno K. Pathways signaling the regulatory volume decrease of cultured nonpigmented ciliary epithelial cells. Invest Ophthalmol Vis Sci 1994; 35: 2876-2886.

23 Lambert IH, Hoffmann EK and Christensen P. Role of protaglandins and leukotrienes in volume regulation by Ehrlich ascites tumor cells. J Membr Biol 1987; 98: 247-256.

24 Jorgensen NK, Lambert IH and Hoffmann EK. Role of LTD_4 in the regulatory volume decrease response in Ehrlich Ascites tumor cells. J Membr Biol 1996; 151: 159-173.

25 Doroshenko P and Neher E. Volume-sensitive chloride conductance in bovine chromaffin cell membrane. J Physiol 1992; 449: 197-218.

26 Hall SK, Zhang J and Lieberman M. Cyclic AMP prevents activation of a swelling-induced chloride-sensitive conductance in chick heart cells. J Physiol 1995; 488: 359-369.

27 Christensen O. Mediation of cell volume regulation by Ca^{2+} influx through stretch-activated channels. Nature 1987; 330: 66-68.

28 Schwiebert EM, Mills JW and Stanton BA. Actin-based cytoskeleton regulates a chloride channel and cell volume in a renal cortical collecting duct cell line. J Biol Chem 1994; 269: 7081-7089.

29 Moran J, Sabanero M, Meza I and Pasantes-Morales H. Changes of actin cytoskeleton during swelling and regulatory volume decrease in cultured astrocytes. Am J Physiol 1996; 271: C1901-C1907.

30 Ingber DE. Tensegrity: the architectural basis of cellular mechanotransduction. Annu Rev Physiol 1997; 59: 575-599.
31 Jennings ML and Schulz RK. Swelling-activated KCl cotransport in rabbit red cells: flux is determined mainly by cell volume rather than shape. Am J Physiol 1990; 259: C960-C967.
32 Thiemann A, Grunder S, Pusch M and Jentsch TJ. A chloride channel widely expressed in epithelial and non-epithelial cells. Nature 1992; 356: 57-60.
33 Grunder S, Thiemann A, Pusch M. and Jentsch TJ. Regions involved in the opening of ClC-2 chloride channel by voltage and cell volume. Nature 1992; 360: 759-762.
34 Jentsch TJ. Chloride channels: a molecular perspective. Curr Opinion Neurobiol 1996; 6: 303-310.
35 Duan D, Winter C, Cowley S, Ume JR and Horowitz B. Molecular identification of a volume-regulated chloride channel. Nature 1997; 390: 417-421.
36 Paulmichl M, Li Y, Wickman K, Ackerman M et al. New mammalian chloride channel identified by expression cloning. Nature 1992; 356: 238-241.
37 Krapivinsky GB, Ackerman MJ, Gordon EA, Krapivinsky LD and Clapham DE. Molecular characterization of a swelling-induced chloride conductance regulatory protein, pI_{Cln}. Cell 1994; 76: 439-448.
38 Ackerman MJ, Wickman KD and Clapham DE. Hypotonicity activates a native chloride current in Xenopus oocytes. J Gen Physiol 1994;10:, 153-179.
39 Voets T, Buyse G, Tytgat J, Droogmans G et al. The chloride current induced by expression of the protein I_{cln} in *Xenopus* oocytes differs from the endogenous volume-sensitive chloride current. J Physiol 1996; 495: 441-447.
40 Jackson PS and Strange K. Single-channel properties of a volume-sensitive anion conductance. Current activation accurs by abrupt switching of channels to an open state. J Gen Physiol 1995; 105: 643-660.
41 Laich A, Furst J, Nagl M, Gschwentner M and Paulmichl M. The transposition of the swelling-activated chloride channel I_{Cln} from the cytosol to the membrane is regulated by the cell volume. J Gen Physiol 1996; 108: 24 (abs).
42 Buyse GT, Voets T, Tytgat J, De Greef C, Droogmans G, Nilius B and Eggermont J. Expresssion of human pI_{Cln} and ClC-6 in Xenopus oocytes induces an identical endogenous chloride conductance. J Biol Chem 1997; 272: 3615-3621.
43 Szucs G, Butse G, Eggermont J, Droogmans G and Nilius B. Characterisation of volume-activated chloride currents in endothelial cells from bovine pulmonary artery. J Membr Biol 1996; 149: 189-197.
44 Valverde MA, Diaz M, Sepulveda FV, Gill DR et al. Volume-regulated chloride channels associated with the human multidrug-resistance P-glycoprotein. Nature 1992; 355: 830-833.
45 Morin XK, Bond TD, Loo TW, Clarke DM and Bear CE. Failure of P-glycoprotein expressed in Xenopus oocytes to produce swelling-activated chloride channel activity. J Physiol 1995; 486: 707-714.
46 Higgins CF. ABC transporters: from microorganisms to man. Annu Rev Cell Biol 1992; 8: 67-113.
47 Valverde MA, Bond TD, Hardy SC, Taylor JC et al. The multidrug resistance P-glycoprotein modulates cell regulatory volume decrease. EMBO J 1996. 15, 4460-4468.

48 Han ES, Vanoye CG, Altenberg GA and Reuss L. P-glycoprotein-associated chloride currents revealed by a specific block by an anti-P-glycoprotein antibody. Am J Physiol 1996; 270: C1370-C1378.

49 Wu J, Zhang JJ, Koppel H and Jacob TJC. P-glycoprotein regulates a volume-activated chloride current in bovine non-pigmented ciliary epithelial cells. J Physiol 1996; 491: 743-755.

50 Zhang JJ, Jacob TJC, Valverde MA, Hardy SP et al. Tamoxifen blocks chloride channels. A possible mechanism for cataract formation. J Clin Invest 1994; 94: 1690-1697.

51 Zhang JJ, Jacob TJC, Higgins CF, Hardy S and Valverde MA et al. Ocular toxicity of anti-oestrogens: tamoxifen vs. ICI 182,780. Br J Pharmacol 1995; 115: 1347-1348.

52 Wang L, Chen L and Jacob TJC. Antisense knock-out of multidrug resistance mRNA (coding for P-gp) reduces magnitude and increases latency of $I_{Cl,vol}$ in boivne epithelial cells. J Physiol 1998 (under review).

53 Luckie DB, Krouse ME, Harper KL, Law TC and Wine JJ.) Selection for MDR1/P-glycoprotein enhances swelling-activated K^+ and Cl^- currents in NIH/3T3 cells. Am J Physiol 1994; 267: C650-C658.

54 Vanoye CG, Altenberg GA and Reuss L. P-glycoprotein is not a swelling-activated Cl^- channel; possible role as a Cl^- channel regulator. J Physiol 1997; 502: 249-258.

55 Miwa A, Ueda K and Okada Y. Protein kinase C-independent correlation between P-glycoprotein expression and volume sensitivity of Cl^- channel. J Membr Biol 1997; 157: 63-69.

56 Pasantes-Morales H, Sanchez Olea R, Miranda D and Moran J.) Volume reulation in NIH/3T3 cells not expressing P-glycoprotein. I. Regulatory volume decrease. Am J Physiol 1997; 272: C1798-C1803.

57 McEwan GTA, Hunter J, Hirst BH and Simmons NL. Volume-activated Cl^- secretion and transepithelial vinblastine secretion mediated by P-glycoprotein are not correlated in cultured T84 intestinal epithelial layers. FEBS Letts 1992; 304: 233-236.

58 Tominaga M, Tominaga T, Miwa A and Okada Y. Volume-sensitive chloride channel activity does not depend on endogenous P-glycoprotein. J Biol Chem 1995; 270: 27887-27893.

59 Luckie DB, Krouse ME, Law TC, Sikic BI and Wine JJ. Doxorubicin selection for MDR1/P-glycoprotein reduces swelling-activated K^+ and Cl^- currents in MES-SA cells. Am J Physiol 1996; 270: C1029-C1036.

60 Weaver JL, Aszalos A and McKinney L. MDR1/P-glycoprotein function. II. Effect of hypotonicity and inhibitors on Cl^- efflux and volume regulation. Am J Physiol 1996; 270: C1453-C1460.

61 Viana F, Van Acker K, De Greef C, Eggermont J et al. Drug-transport and volume-activated chloride channel functions in human erythroleukemia cells: relation to expression level of P-glycoprotein. J Memb Biol 1995; 145: 87-98.

62 De Greef C, van der Hayden S, Viana F, Eggermont J, et al. Lack of correlation between mdr-1 expression and volume-activation of chloride-currents in rat colon cancer cells. Pflugers Arch 1995; 430: 296-298.

63 Altenberg GA, Vanoye CG, Han ES, Deitmer JW and Reuss L. Relationships between rhodamine 123 transport, cell volume, and ion-channel function of P-glycoprotein. J Biol Chem 1994; 269: 7145-7149.

64 Rasola A, Galietta LJV, Gruenert DC and Romeo G. Volume-sensitive chloride currents in four epithelial cell lines are not directly correlated to the expression of the MDR-1 gene. J Biol Chem 1994; 269: 1432-1436.

65 Ehring GR, Osipchuk YV and Cahalan MD. Swelling-activated chloride channels in multidrug-sensitive and -resistant cells. J Gen Physiol 1994; 104: 1129-1161.

66 Mitchell CH, Zhang JJ, Wang L and Jacob TJC. Volume-sensitive chloride current in pigmented ciliary epithelial cells: role of phospholipases. Am J Physiol 1997; 272: C212-C222.

67 Zhang JJ and Jacob TJC. Three different Cl^- channels in the bovine ciliary epithelium activated by hypotonic stress. J Physiol 1997; 499: 379-389.

68 Anderson MP, Gregory RJ, Thompson, Souza DW et al. Demonstration that CFTR is a chloride channel by alteration of its anion selectivity. Science 1991; 253: 202-205.

69 McNicholas CM, Guggino WB, Schwiebert EM, Nebert SC et al. Sensitivity of a renal K^+ channel (ROMK2) to the inhibitory sulfonylurea compound glibenclamide is enhanced by coexpression with the ATP-binding cassette transporter cystic fibrosis transmembrane regulator. PNAS (USA) 1996; 93: 8083-8088.

70 Mall M, Kunzelmann K, Hipper A, Busch AE and Greger R. cAMP stimulation of CFTR-expressing Xenopus oocytes activates a chromanol-inhibitable K^+ conductance. Pflugers Arch 1996; 432: 516-522.

71 Takano M, Ishii T and Xie L-H. Cloning and functional expression of the rat brain KIR6.2 channel. Jap J Physiol 1996; 46: 491-495.

72 Okada Y. Volume expansion-sensing outward-rectifier Cl^- channel: fresh start to the molecular identity and volume sensor. Am J Physiol 197;273: C755-C789.

73 Kirk K. Swelling-activated organic osmolyte channels. J Membr Biol 1997;158: 1-16.

74 Thomas PM, Cote GJ, Wohllk N, Haddad B et al. Mutations in the sulphonylurea receptor gene in familial persistent hyperinsulinemic hypoglycemia of infancy. Science 1995; 268: 426-429.

75 Aguilarbryan L, Nichols CG, Wechsler SW, Clement JP, et al. Cloning of the beta-cell high-affinity sulphonylurea receptor - a regulator of insulin secretion. Science 1995; 268: 423-426.

76 Zhang JJ and Jacob TJC. ATP-activated chloride channel inhibited by an antibody to P glycoprotein. Am J Physiol 1994; 267: C1095-C1102.

77 Reymann S, Florke H, Heiden M, Jakib C et al. Further evidence for multitopological localization of mammalian porin (VDAC) in the plasmalemma forming part of a chloride channel complex affected in cystic fibrosis and encephalomyopathy. Biochemical and Molecular Medicine 1995; 54: 75-87.

78 Moorman JR, Ackerman SJ, Kowdley GC, Groffin MP et al. Nature 1995; 377: 737-740.

79 Kowdley GC, Ackerman SJ, Chen Z Szabo G et al. Anion, cation, and zwitterion selectivity of phospholemman channel molecules. Biophys J 1997; 72: 141-145.

80 Morrison BW, Moorman JR, Kowdley GC, Kobayashi YM et al. Mat-8, a novel phospholemman-like protein expressed in human breast tumor, induces a chloride conductance in Xenopus oocytes. J Biol Chem 1995; 270: 2176-2182.

81 Goldstein L, Brill SR and Freund EV. Activation of taurine efflux in hypotonically stressed elasmobranch cells: inhibition by stilbene disulphonates. J Exp Zool 1990; 254: 114-118.

82 Goldstein L and Brill SR. Volume-activated taurine efflux from skate erythrocytes: possible *band 3* involvement. Am J Physiol 1991; 260: R1014-R1020.

83 van Es HHG, Renkma H, Aerts H and Schurr E. Mol Biochem Pharmacol 1994; 68: 209-219.

84 Higgins CF. The ABC of channel regulation. Cell 1995; 82: 693-696.

85 Gschwentner M, Nagl UO, Woll E, Schmarda A et al. Antisense oligonucleotides suppress cell-volume-induced activation of chloride channels. Pflugers Arch 1995; 430: 464-470.

86 Nagl UO, Erdel M, Schmarda A, Seri M et al. Chromosomal localization of the genes (CLNS1A and CLNS1B) coding for the swelling-dependent chloride channel ICln. Genomics 1996; 38:438-441.

87 Attali B, Guillemare E, Lesage F, Honore, et al. The protein IsK is a dual activator of K^+ and Cl^- channels. Nature 1993; 365: 850-852.

88 Banderali U and Ehrenfeld J. Heterogeneity of volume-sensitive chloride channels in basolateral membranes of A6 epithelial cells in culture. J Membr Biol 1996;154: 23-33.

Cell Volume Regulation: The Molecular Mechanism and Volume Sensing Machinery
Y. Okada, editor.

Electrophysiological properties of volume-regulated Cl^- channels in intestinal epithelial cells

Shigetoshi Oiki, Machiko Kubo, Yasunobu Okada

Department of Cellular and Molecular Physiology, National Institute for Physiological Sciences, Okazaki 444-8585, Japan

SUMMARY

Regulation of their volume is a fundamental homeostatic function of animal cells. In a wide variety of cell types, the regulatory volume decrease which follows osmotic swelling is accomplished by parallel activation of K^+ and Cl^- conductances followed by obligatory water efflux. Depletion of cytosolic ATP abolished swelling-induced activation of Cl^- currents in human epithelial cells (Intestine 407 cells). The ATP role could be substituted by AMP-PNP, ATPγS, GTP and GTPγS. Blockers for protein kinases and protein phosphatases failed to affect swelling-activated Cl^- currents. Thus, it is concluded that non-hydrolytic binding of ATP is involved in regulation of the volume-sensitive Cl^- channel. The regulation by cytosolic ATP of the volume-sensitive Cl^- channel may provide a fine tuning mechanism for cell volume regulation under osmotic or metabolic perturbation.

PHENOTYPIC PROPERTIES OF EPITHELIAL VOLUME-SENSITIVE Cl^- CHANNELS

The regulatory volume decrease (RVD) is known to be attained by parallel activation of K^+ and Cl^- channels in a variety of cell types [1,2]. We have studied biophysical/physiological/ pharmacological properties of the volume-sensitive Cl^- channel in human epithelial Intestine 407 cells [3-9]. Table 1 summarizes the phenotypic properties. Among them, dependence of the channel activity on cytosolic free ATP [4] is most physiologically important.

DEPENDENCE ON CYTOSOLIC ATP BINDING

When voltage-clamped human epithelial Intestine 407 cells were exposed to a hypotonic solution (83% osmolality) in the whole-cell mode, activation of Cl^- channel currents took place in association with osmotic swelling. Figure 1 (*upper traces*) represents the typical Cl^- currents recorded following application of alternating positive and negative voltage pulses in the steady state with a patch pipette containing 0.1 mM free ATP and 1 mM free Mg^{2+}. The Cl^- current exhibited time- and voltage-dependent inactivation upon application of pulses positive to +40 mV. When intracellular ATP was depleted by perfusing the cytosol with an ATP-free pipette solution containing metabolic inhibitors (FCCP, iodoacetic acid *plus* oligomycin) and 1 mM free Mg^{2+}, as observed in our previous study [4], hypotonic challenge failed to induce activation of Cl^- currents. A similar requirement for ATP has been noted for volume-sensitive Cl^- currents in other cell types (see Refs. in recent reviews [1,2]).

Table 1
Summary of properties of volume-sensitive Cl^- channel in Intestine 407 cells.

single-channel	intermediate unitary conductance [6,7] (~50 pS at +100 mV, ~20 pS at −100 mV) high open probability (~0.97 at −100 mV) [6,7]
voltage dependence	moderate outward rectification [3,6,7] faster inactivation at larger positive potentials [3,6,7] no activation and inactivation kinetics at negative potentials [3,6,7]
anion selectivity	$SCN^- > I^- > Br^- > Cl^- > F^- >$ gluconate$^-$ [3] (low field selectivity: Eisenman sequence I)
ATP dependence	cytosolic free ATP requirement [3] extracellular ATP blocking action [7]
Mg^{2+} dependence	cytosolic free Mg^{2+} sensitivity [7]
pharmacology	broad sensitivity to Cl^- channel blockers [1,3] arachidonic acid sensitivity [3] sensitivity to some P-glycoprotein blockers [8] glibenclamide sensitivity [9] no sensitivity to blokers for protein kinases and phosphatases (shown in this report)

Under conditions of constant $[Mg^{2+}]_i$, the Cl^- currents increased with increasing the intracellular concentration of free ATP ($[ATP]_i$). The $[ATP]_i$ for half-maximal activation (EC_{50}) was very low (0.011 mM at 1 mM free Mg^{2+}). Even when intracellular Mg^{2+} ions were largely chelated by EDTA (5 mM) in the presence of intracellular ATP, the volume-sensitive Cl^- currents were not attenuated (Fig. 2: $[Mg^{2+}]_i = 10^{-5}$ mM). Therefore, it appears that the binding of free ATP, not MgATP, activates the channel.

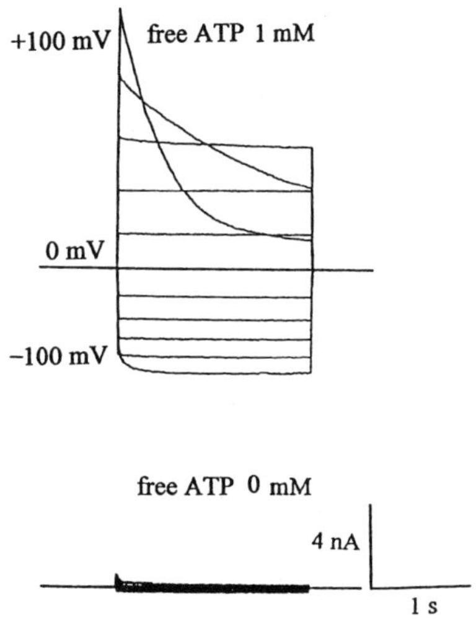

Fig. 1. Whole-cell Cl^- current responses to positive (+20 to +100 mV) and negative (−20 to −100 mV) pulses applied alternatively to Intestine 407 cells equilibrated with CsCl solution. Intracellular ATP was depleted by following metabolic inhibitors that were contained in a patch pipette: FCCP (1 μM), iodoacetic acid (1 μM) and oligomycin (40 μM).

INDEPENDENCE OF PHOSPHORYLATION / DEPHOSPHORYLATION

A slowly hydrolyzable ATP analogue, ATPγS, and a non-hydrolyzable ATP analogue, AMP-PNP, did not abolish but could rather substitute, at least in part, for ATP (Fig. 2), as observed in other cell species [1,2]. Thus, the role of free ATP is not mediated by its hydrolysis, but rather by its binding to some functional component of the volume-sensitive Cl^- channel. The ATP role could be, at least in part, substituted by a variety of nucleotides including dATP, ADP, GTP, GTPγS, TTP, ITP, UTP and CTP, as shown in Fig. 2.

A high concentration of the protein kinase blocker H-8 (100 μM), which may inhibit almost all types of protein kinase, failed to inhibit activation of the Cl^- currents (Fig. 2, Inset). Phosphatase blockers (okadaic acid 5 μM, fluoride 10 mM or Na-vanadate 0.5 mM) were not able to inhibit the volume-regulated Cl^- currents (Fig. 2, Inset). Therefore, neither phosphorylation nor dephosphorylation played any significant role in activation the channel.

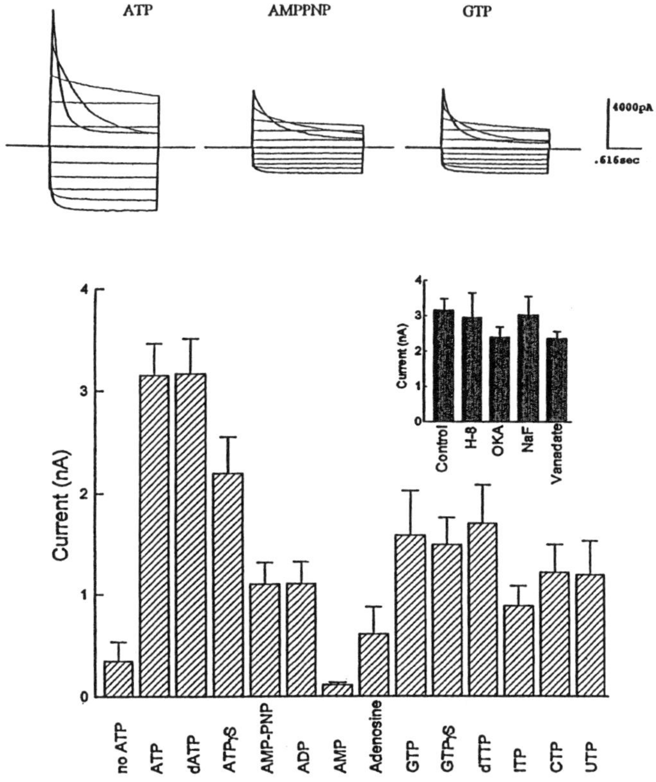

Fig. 2. Effect of various nucleotides and related substances on the swelling activated Cl^- channel. The whole-cell current amplitudes at +40 mV after attaining steady-state current level under persistent hypotonic stress are shown. The concentration of each free nucleotide was 0.1 mM. Inset: Effects of blockers for protein phosphatases and kinases.
n = 4 ~ 6.

REFERENCES

1 Okada Y. Volume expansion-sensing outward-rectifier Cl^- channel: fresh start to the molecular identity and volume sonsor. Am J. Physiol. 1997; 273: C755-C789.

2 Strange K, Emma F, Jackson P S. Cellular and molecular physiology of volume-sensitive anion channels. Am. J. Physiol. 1996; 270: C711-C730.

3 Kubo M, Okada Y. Volume-regulatory Cl^- channel currents in cultured human epithelial cells. J. Physiol. 1992; 456: 351-371.

4 Oiki S, Kubo M, Okada Y. Mg^{2+} and ATP-dependence of volume-sensitive Cl^- channels in human epithelial cells. Jpn. J. Physiol. 1994; 44: S77-S79.

5 Okada, Kubo M, Oiki S, Petersen CCH, Tominaga M, Hazama A Morishima S. Properties of volume-sensitive Cl^- channels in a human epithelial cell line. Jpn. J. Physiol. 1994; 44: S31-S35.

6 Okada Y, Petersen CCH, Kubo M, Morishima S, Tominaga M. Osmotic swelling activates intermediate-conductance Cl^- channels in human intestinal epithelial cells. Jpn. J. Physiol. 1994; 44: 403-409.

7 Tsumura T, Oiki S, Ueda S, Okuma M, Okada Y. Sensitivity of volume-sensitive Cl^- conductance in human epithelial cells to extracellular nucleotides. Am. J. Physiol. 1996; 271: C1872-C1878.

8 Tominaga M, Tominaga T, Miwa A, Okada Y. Volume-sensitive chloride channel activity does not depend on endogenous P-glycoprotein. J. Biol. Chem. 1995; 270: 27887-27893.

9 Liu Y, Tumura T, Oiki S, Okada Y. Glibenclamide inhibits volume-sensitive Cl^- cunductance in human epithelial cells. Jpn. J. Physiol. 1997; 47: Suppl. 2: S123.

Cell Volume Regulation: The Molecular Mechanism and Volume Sensing Machinery
Y. Okada, editor.

Is ICln a swelling dependent chloride channel?

M. Gschwentner[a], J. Fürst[a], A. Schmarda[a], C. Bazzini[a], E. Wöll[b], M. Ritter[b], A. Dienstl[b], E. Scandella[a], B. Oehl[a], M. Hohlrieder[a] and M. Paulmichl[a]

[a] Dept. of Physiology, University of Innsbruck, Fritz-Pregl Strasse 3, A-6020 Innsbruck, Austria

[b] Dept. of Internal Medicine, University of Innsbruck, Anich Strasse 35, A-6020 Innsbruck, Austria

SUMMARY

Since the cloning of ICln in 1992 a considerable number of experiments has been published to elucidate the function of this protein. As summarized here our hypothesis of ICln being the swelling dependent chloride channel or a substantial part thereof is supported by the experimental data. However, more experiments need to be done to determine the precise function of ICln in regulatory volume decrease (RVD).

INTRODUCTION

It is commonly believed that the activation of chloride and potassium channels after cell swelling is a key strategy for cells to reduce their volume, thus readjusting their size to the pre-swelling level [1-3]. The cell swelling itself can be effected by several means, such as sudden reduction of osmolarity in the extracellular fluid or increase of osmolarity in the cytosol. The reduction of extracellular osmolarity can be caused by an absolute reduction of osmolytes (this is a standard approach used for most studies summarized here), or by first increasing and then reducing osmolarity. Both forms of osmolarity reduction are encountered in the living organism. Another way to increase cytoplasmatic volume is to increase cytoplasmatic osmolarity which leads to a concomitant water flux into the cells driven by the osmotic gradient. Such events can be caused by substantial ion or solute (glucose, amino acids etc.) uptake due to stimulated transport mechanisms or intracellular catabolism. Since the function of organs is usually tightly linked to the cytoarchitecture, the control of cell volume is vital for the organism.

As mentioned above, one strategy for cells to effect the described volume regulation after swelling is the activation of ion channels. It is of utmost importance to define those channels on a molecular level, since this is the prerequisite to investigate their structure/function relation. No convincing candidate protein has so far been isolated to be envisaged as the cation pathway involved in regulatory volume decrease. However, several different proteins were described in the past as molecular candidates for the anion channel activated during RVD. The list of those proteins comprises structures like the multidrug transporter mdr1 or p-glycoprotein [4, 5], ClC-2 and 3 [6, 7], phospholemman [8], NCC27 [9] and ICln [10]. All those proteins were originally identified by overexpression in *Xenopus laevis* oocytes or mammalian cells where they create anion currents similar to the ones encountered in cells solely subjected to swelling. The similarity of the induced currents after expressing the respective proteins with the swelling dependent chloride currents leads to their linkage to RVD and sparks speculations on their function being the channels responsible for RVD. There is, however, an inherent problem common to all experiments based on the overexpression of putative swelling dependent chloride channels, namely the fact that *no* cell is known with a proven absence of swelling dependent chloride channels. Therefore, expressing putative swelling dependent chloride channels in cell systems inevitably means that putative swelling dependent chloride channels are expressed *on top* of endogenously existing ones. This circumstance makes it obvious that, based on such experiments, it *cannot* be unambiguously decided if the expressed proteins are the channels themselves or modifiers of endogenously present channels.

In the following section we summarize experiments which led us to the assumption that ICln is most likely a swelling dependent chloride channel or a significant part of it.

RESULTS AND DISCUSSION

ICln was originally cloned from Madin Darby canine kidney (MDCK)-cells [10]. The name ICln was given to the protein for inducing a nucleotide sensitive chloride current when expressed in *Xenopus laevis* oocytes (I=current, Cl=chloride and n=nucleotide sensitive). Consequently, the ICln protein was also termed pICln [11], however, since this abbreviation was used in the original cloning paper for the plasmid bearing the ICln clone, it should *not* be applied to the ICln protein. The expression of ICln in *Xenopus laevis* oocytes is followed by the development of a chloride current evidenced by the reversal potential and the ability of NPPB and DIDS to block the expressed current. The ICln induced current in oocytes is, furthermore, outwardly rectifying and inactivated at voltages more positive than 40 mV. An important feature of ICln expressed in oocytes is the ability of nucleotides added to the extracellular fluid to block the induced current. A substantial reduction of the ICln current can be observed with cAMP, ITP, cGMP and GTP at a concentration of 1 mM, whereas ATP, ADP, AMP and adenosine require an approximately 10 times higher concentration to cause the same effect.

According to the chloride channel in the outer membrane of mitochondria (VDAC; [12]) ICln was modeled as a ß-barrel consisting of two ICln units. In this model antiparallel ß-strands cross the inner and outer leaflet of the membrane and a cluster of glycines (G49, G51 and G53) forms part of the extracellular entrance of the pore. Similar glycine clusters were described as P-loops and identified in an increasing group of proteins as the ATP/GTP binding site [13]. The mutation of the glycine at position 49 (G49) in the ICln protein was followed by a substantial reduction of the ability of cAMP to block the induced current showing that this glycine is involved in the binding of cAMP at the ICln protein and, furthermore, that this binding site is accessible from the extracellular side. The simplest conclusion from those experiments is to envisage ICln as the channel itself or a substantial part of it. However, as mentioned before [10], more complex interpretations are feasible. Since *Xenopus laevis* oocytes bear endogenous swelling dependent chloride channels [14, 15] which, activated by a reduction of extracellular osmolarity by 50%, lead to a current similar to the current observed after the expression of ICln. Hence, the possibility must be considered that ICln could also be a regulator of the endogenously present swelling dependent chloride channels. It is important to mention that the current measured in oocytes after the expression of ICln is observed under isotonic conditions [10, 16-19], whereas the endogenous swelling dependent chloride current in oocytes is observed after the reduction of extracellular osmolarity [14, 15]. A possible explanation for this is that due to the immense overexpression of ICln, potential negative controlling mechanisms in the cells are overpowered and thus the active channel can be seen even in the absence of the appropriate stimulus (it is a consistent finding that under isotonic conditions swelling dependent chloride currents can hardly be seen [1, 2, 20-25]). Nilius et al. described a slight, but statistically significant difference between the current evoked by the expression of human ICln in oocytes and the endogenous swelling dependent chloride current in *Xenopus laevis* oocytes [19]. Whereas the authors claim a difference in the rectification which is not supported by our data or data obtained by others [14, 15], the different amino acid sequences within the predicted pore forming units of human and frog ICln could be made responsible for all the other minor deviations described involving functions closely related to the channel pore. Five of the seven amino acids which are different in this area in the two species are predicted to face the pore. According to our hypothesis of ICln being the channel it would be very interesting to see if, by using site directed mutagenesis, those amino acids are related to the observed differences. On the other hand, these experiments do not prove the hypothesis that ICln activates endogenous swelling dependent chloride channels, since it is clearly shown that the expression of ICln is not compellingly linked to such an activation. According to these experiments ICln could, however, be envisaged as activating *not yet* identified chloride channels in oocytes. It was also reported that ClC-6 leads to a current similar to the current induced by ICln [18], but, Jentsch et al. were not able to see this current after the expression of ClC-6 [26]. Furthermore it was reported that the expression of rat ICln bearing the same mutations we used to identify the putative nucleotide binding site (see above) leads to a current in oocytes which cannot clearly be distinguished from the endogenous swelling dependent chloride current in the same cells [27]. These experiments substantiate the question under which conditions the expression of foreign proteins in *Xenopus leavis* oocytes can activate the native swelling dependent chloride channels.

Moreover, they point inevitably towards the dilemma that the expression of swelling dependent chloride channel candidates in a system bearing closely related channels is not able to give unambiguous answers to the question about the channel-function of those proteins.

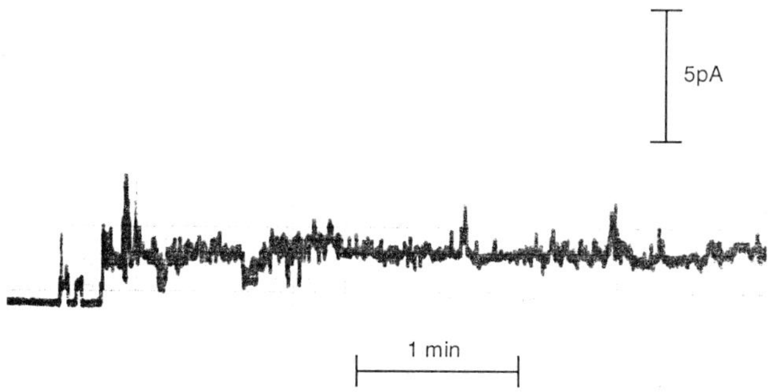

Figure 1: Functional reconstitution of ICln into black lipid membrane by adding ≈ 0.6 μg purified protein to the salt solution of the *cis* side of the setup (the solutions used are composed of (in mM) 100 KCl, 5 HEPES (pH 8.0) and the holding potential was -100 mV).

In order to test a possible link between ICln and swelling dependent chloride channels we used a transient knock-down approach which allowed us to avoid the overexpression of ICln for the reasons mentioned. The idea was that if ICln is related to swelling dependent chloride channels, the substantial reduction of ICln in native cells should influence the swelling dependent chloride current in these cells. The knock-down of ICln in NIH 3T3 fibroblasts was carried out by antisense oligodeoxynucleotides complementary to the initial part of the ORF coding for ICln [28]. In these experiments we could indeed show that the reduction of ICln in fibroblasts is followed by a substantial reduction of the swelling dependent chloride current. Although these data show that ICln and swelling dependent chloride channels are clearly linked in fibroblasts they provide no further insight into the question of whether ICln is a channel or not.

A significant amount of the ICln protein can be identified in the water soluble fraction of the cytosol of undifferentiated cells. This was described by us [29] and others [11, 18]. Under isotonic conditions only a minor part of the total amount of ICln in the cells can be identified in a close relation with the plasma membrane, however, this part can be significantly increased by challenging the cells with a reduction of extracellular osmolarity by 50% or more [29, 30]. A 40% reduction is obviously not enough for the translocation of ICln in EAhy926 cells [18]. We speculated that similar to bacterial toxins [31] the ICln protein could be transposed directly from the cytosol into the membrane where it forms active channels [20, 24, 29, 32-35]. The same hypothesis was also discussed by others based on experiments using noise analysis in C6 glioma cells [21, 36]. The transposition hypothesis as well as the channel hypothesis for ICln can be tested by reconstituting the ICln protein into black lipid bilayers. As shown in **figure 1** the addition of purified ICln protein to the solution in the *cis* compartment of the bilayer setup led to the development of channel like currents, showing that ICln can act as a channel and a direct transposition of the protein from the aqueous phase into the membrane can occur.

However, the reconstitution experiments have so far failed to disclose the exact matching current phenotype encountered in native cells after swelling. Since ICln is clearly linked to RVD, further experiments are necessary to reveal the conditions (ion composition of the solutions used and/or additional 'channel-proteins' in order to allow the formation of hetero-multimeres) under which these experiments match the experiments in native cells.

Table 1
ICln facts:

1.	Nomenclature	ICln
2.	Other names	pICln, I_{Cln}
3.	Amino acids (MDCK)	235
4.	Gene name and chromosomal localization (human)	CLNS1A, 11q13.5-14.1 and CLNS1B, 6p12[a]
5.	Homologues and GeneBank access#	MDCK X65450, human U17899, rabbit D26076, rat D13985/L26450, X. laevis L26449 and zebra fish Y08484[b]
6.	Tissue expression	skeletal muscle, heart, brain, lung, kidney, pancreas, stomach, intestine, liver, NIH 3T3, PC12, endothelia and many others[c]
7.	Current rectification	outward
8.	Reconstitution in bilayer	yes
9.	Permeabilities	$SCN^- > I^- > Br^- > Cl^-$
10.	Blockers	DIDS, SITS, NPPB, 9-AC, nucleotides[d]
11.	Regulation	Inactivation at positive potentials
12.	Structural information	Probable multiunit ß-barrel
13.	Interaction with other proteins	actin[e]

References: a [37, 38], b [10, 14, 16, 17, 39, 40], c [10, 16, 17, 38, 39, 41], d [10, 18, 19, 42], e [11, 43].

CONCLUSION

The experiments summarized here strongly support our hypothesis of ICln being a channel or a significant part thereof. Extensive additional studies need to be done to unambiguously determine the role of ICln in RVD.

ACKNOWLEDGEMENTS

The expert technical assistance by G. Buemberger, E. Papp, A. Wimmer and P. Rouhi is gratefully acknowledged. This work was supported in part by grants from the Austrian Science Foundation, the Union Bank of Switzerland, the Austrian National Bank, the Gastein Foundation, BIOMED and the Rockefeller Foundation to MP (markus.paulmichl@uibk.ac.at).

REFERENCES

1. Hoffmann, E.K. and L.O. Simonsen, *Membrane mechanisms in volume and pH regulation in vertebrate cells.* Physiological Reviews, 1989. **69**: p. 315-382.
2. Lang, F., H. Völkl, and D. Häussinger, *General principles in cell volume regulation.* Comparative Physiology, 1990. **4**: p. 1-25.
3. Lang, F., H. Völkl, and M. Paulmichl, *How do cells regulate their volume.* Pflügers Archiv, 1988. **411**: p. R4-R5.
4. Valverde, M.A., *et al.*, *Volume-regulated chloride channels associated with the human multidrug-resistance P-glycoprotein.* Nature, 1992. **355**: p. 830-833.
5. Gill, D.R., *et al.*, *Separation of drug transport and chloride channel functions of the human multidrug resistance P-glycoprotein.* Cell, 1992. **71**: p. 23-32.
6. Duan, D., *et al.*, *Molecular identification of a volume-regulated chloride channel.* Nature, 1997. **390**: p. 417-421.
7. Gründer, S., *et al.*, *Regions involved in the opening of ClC-2 chloride channel by voltage and cell volume.* Nature, 1992. **360**: p. 759-762.
8. Moorman, R.J., *et al.*, *Unitary anion currents through phospholemman channel molecules.* Nature, 1995. **377**: p. 737-740.
9. Valenzuela, S.M., *et al.*, *Molecular Cloning and Expression of a chloride channel of Cell Nuclei.* Journal of Biological Chemistry, 1997. **272**: p. 12575-12582.
10. Paulmichl, M., *et al.*, *New mammalian chloride channel identified by expression cloning.* Nature, 1992. **356**: p. 238-241.

11. Krapivinsky, G.B., *et al.*, *Molecular characterization of a swelling-induced chloride conductance regulatory protein, pI_{Cln}.* Cell, 1994. **76**: p. 439-448.
12. Forte, M., D. Adelsberger-Mangan, and M. Colombini, *Purification and characterization of the voltage-dependent anion channel from the outer mitochondrial membrane of yeast.* Journal of Membrane Biology, 1987. **99**: p. 65-72.
13. Saraste, M., P.R. Sibbald, and A. Wittinghofer, *The P-loop - a common motif in ATP- and GTP-binding proteins.* Trends in Biological Sciences, 1990. **15**: p. 430-434.
14. Ackerman, M.J., K.D. Wickman, and D.E. Clapham, *Hypotonicity activates a native chloride current in Xenopus oocytes.* Journal of General Physiology, 1994. **103**: p. 153-179.
15. Hand, M., R. Morrison, and K. Strange, *Characterization of Volume-sensitive Organic Osmolyte Efflux and Anion Current in Xenopus Oocytes.* Journal Membrane Biology, 1997. **157**: p. 9-16.
16. Okada, H., *et al.*, *Cloning of a swelling-induced chloride current related protein from rabbit heart.* Biochemica et Biophysica Acta, 1995. **1234**: p. 145-148.
17. Abe, T., *et al.*, *Molecular cloning and expression of a rat cDNA encoding MDCK-type chloride channel.* Biochimica et Biophysica Acta, 1993. **1173**: p. 353-356.
18. Buyse, G., *et al.*, *Expression of human pICln and CLC-6 in Xenopus oocytes induces an identical endogenous chloride conductance.* Journal of Biological Chemistry, 1997. **272**: p. 3615-3621.
19. Voets, T., *et al.*, *The chloride current induced by expression of the protein pICln in Xenopus oocytes differs from the endogenous volume-sensitive chloride current.* Journal of Physiology, 1996. **495**: p. 441-447.
20. Gschwentner, M., *et al.*, *ICln a chloride channel paramount for cell volume regulation.* Journal of Allergy and Clinical Immunology, 1996. **98**: p. S98-S101.
21. Jackson, P.S. and K. Strange, *Single-channel properties of a volume-sensitive anion conductance. Current activation occurs by abrupt switching of closed channels to an open state.* Journal of General Physiology, 1995. **105**: p. 643-660.
22. Nilius, B., J. Sehrer, and G. Droogmans, *Permeation properties and modulation of volume-activated Cl^--currents in human endothelial cells.* British Journal of Pharmacology, 1994. **112**: p. 1049-1056.
23. Okada, Y., *Volume expansion-sensing outward-rectifying Cl- channel: fresh start to the molecular identity and volume sensor.* American Journal of Physiology, 1997. **273**: p. C755-C789.
24. Paulmichl, M., A. Norris, and D. Rainey, *Role of chloride channel modulation in the mechanism of action of nedocromil sodium.* International Archives of Allergy and Clinical Immunology, 1995. **107**: p. 416.
25. Strange, K., F. Emma, and S.P. Jackson, *Cellular and molecular physiology of volume-sensitive anion channels.* American Journal of Physiology, 1996. **270**: p. C711-C730.

26. Brandt, S. and T. Jentsch, *ClC-6 and ClC-7 are two novel broadly expressed members of the CLC chloride channel family*. FEBS Letters, 1995. **377**: p. 15-20.
27. Voets, T., *et al.*, *Mutation of the putative nucleotide-binding site of pICln does not alter the nucleotide sensitivity and Ca++ dependence of the chloride current ICln after expression in Xenopus oocytes.* J Physiology, 1998. **506P**: p. 47P.
28. Gschwentner, M., *et al.*, *Antisense oligonucleotides suppress cell-volume-induced activation of chloride channels.* Pflügers Archiv, 1995. **430**: p. 464-470.
29. Laich, A., *et al.*, *The transposition of the swelling-dependent chloride channel I_{Cln} from the cytosol to the membrane is regulated by the cell volume.* Journal of General Physiology, 1996. **in press**.
30. Musch, M.W., *et al.*, *Hypotonic Stress Induces Translocation of the Osmolyte Channel Protein pICln in Embrionic Skate (Raja eglanteria) Heart.* Journal of Experimental Zoology, 1997. **277**: p. 460-463.
31. Parker, M.W., *et al.*, *Structure of the membrane-pore-forming fragment of colicin A.* Nature, 1989. **337**: p. 93-96.
32. Gschwentner, M., *et al.*, *Further characterization of a cloned epithelial chloride channel and its role in cell volume regulation (RVD).* Pflügers Archiv, 1993. **422**: p. R64.
33. Gschwentner, M., *et al.*, *Structure-function relation of a cloned epithelial chloride channel.* Renal Physiology and Biochemistry, 1994. **17**: p. 148-152.
34. Laich, A., *et al.*, *ICln, a chloride channel cloned from kidney cells, is activated during regulatory volume decrease.* Kidney International, 1997. **51**.
35. Paulmichl, M., *et al.*, *Insight into the Structure-Function of Chloride Channels.* Cellular Physiology and Biochemistry, 1993(3): p. 374-387.
36. Jackson, P.S. and K. Strange, *Characterization of the voltage-dependent properties of a volume-sensitive anion conductance.* Journal of General Physiology, 1995. **105**: p. 661-677.
37. Nagl, U., *et al.*, *Characterization of the human Gene Coding for the Swelling-Dependent Chloride Channel ICln at Position 11q13.5-14.1 (CLNS1A) and further Characterization of the Chromosome 6 (CLNS1B) Localization.* Gene, 1998. **in press**.
38. Nagl, O.U., *et al.*, *Chromosomal Localization of the Gene Coding for the Swelling-Dependent Chloride Channel ICln.* Genomics, 1996. **38**: p. 438-441.
39. Anguita, J., *et al.*, *Molecular Cloning of the Human Volume-sensitive Chloride conductance regulatory protein pICln from Ocular Ciliary Epithelium.* Biochemical and Biophysical Research Communications, 1995. **208**: p. 89-95.
40. Schmarda, A., *et al.*, *Cloning of the swelling-dependent Chloride Channel ICln Homologue from Zebra Fish.* Cell Physiology and Biochemistry, 1997. **7**: p. 298-302.
41. Buyse, G., *et al.*, *The ubiquitously Expressed pICln Protein Forms Homomeric Complexes in Vitro.* Biochemical and Biophysical Research Communications, 1996. **218**: p. 822-827.

42. Gschwentner, M., *et al.*, *Antiviral drugs from the nucleoside analog family block volume-activated chloride channels.* Molecular Medicine, 1995. **1**: p. 407-417.
43. Schwarz, R.S., A.C. Rybicki, and R.L. Nagel, *Molecular cloning and expression of a chloride channel-associated protein pICln in human young red blood cells: association with actin.* Biochem. J., 1997. **327**: p. 609-616.

Cell Volume Regulation: The Molecular Mechanism and Volume Sensing Machinery
Y. Okada, editor.

Volume-regulated anion channels in vascular endothelium

B. Nilius, J. Eggermont, T. Voets, and G. Droogmans

KU Leuven, Campus Gasthuisberg, Department of Physiology, B-3000 LEUVEN, Belgium

SUMMARY

Volume-regulated anion channels (VRAC) open in response to an increase in cell volume are abundantly expressed in most excitable and non-excitable cells. The corresponding current is under physiological conditions mainly carried by Cl^- anions. Amino acids and organic osmolytes also permeate through the volume-sensitive anion channel. The swelling-activated efflux of Cl^- and organic osmolytes is functionally important for the regulation of cell volume. In addition, VRAC may also contributes to control of membrane potential, vectorial transport of solutes, Ca^{2+} influx, and intracellular pH homeostasis and could assist in endothelial cells regulating cell growth and the transition from proliferation to differentiation.

This article focuses on volume-regulated anion currents in vascular endothelial cells, which is outwardly rectifying, inactivates at positive potentials and is more permeable for iodide than for chloride. Permeability for organic osmolytes and some amino acids such as taurine is surprisingly high. We discuss the molecular biophysical features of VRAC, permeation properties and gating mechanisms.

The mechano-transduction cascade responsible for activation of VRAC is not understood. It will be approached by the likely involvement of tyrosine phospholylation cascades including src and rho and changes in intracellular ionic strength.

The classical Cl^- channels blockers, such as DIDS, NPPB, 9-AC, NPA, inhibit VRAC with low affinity. Compounds such as the non-charged quinine or quinidine, the antiestrogen tamoxifen, and phenol derivatives, such as gossypol, are more sensitive blockers.

The present data do not offer a clear picture of the molecular identity of the channel nor of the molecular composition of the regulatory cascade.

INTRODUCTION

As in many - if not all - excitable and non-excitable cells, vascular endothelial cells abundantly express anion channels that open in response to an increase in cell volume. These volume-regulated anion channels (VRAC) carry under physiological conditions mainly Cl^- anions but also amino acids and organic osmolytes permeate through it. The channel is also referred to as VSOAC (volume-sensitive organic osmolyte anion channel) [for reviews see 1, 2, 3]. In endothelial cells, VRAC is already activated under resting conditions and is functionally important for regulation of the membrane potential of these cells. It contributes vectorial transport of solutes, regulation of Ca^{2+} influx, and intracellular pH homeostasis. Furthermore, it could assist in regulating cell growth and the transition from proliferation to differentiation.

VOLUME REGULATED ANION CHANNELS IN ENDOTHELIUM

We have described for the first time VRAC in vascular endothelial cells [4].

Molecular biology: In addressing the molecular identity of VRAC we have focused on MDR1 P-glycoprotein and pI_{Cln}. P-glycoprotein (P-gp) is expressed in endothelial cells, but the lack of correlation between P-gp expression and density of VRAC as well as the absence of a mutual interference between P-gp transport activity and VRAC argue against P-gp being VRAC [5-7]. We also addressed the relation between the pI_{Cln} protein and VRAC. To this end, we cloned human pI_{Cln} and studied its expression pattern at the mRNA and protein level [8]. Expression studies in Xenopus oocytes clearly revealed that the current induced by expression of human pI_{Cln} differs from the endogenous VRAC [9]. Moreover, there was no shift of pI_{Cln}, which is mainly cytosolic, to the plasma membrane if cells were exposed to a hypotonic solution [10]. Finally, the anion current observed after expression of pI_{Cln} in Xenopus oocytes corresponds to an endogenous current [10]. We have also performed RT-PCRs with degenerated oligonucleotide primers to test whether VRAC may be encoded by a member of the ClC family. This search revealed tissue-dependent alternative splicing of ClC-6 transcripts [11], but we obtained no positive evidence for VRAC being a ClC-type channel. To our conclusion, the molecular nature of VRAC is not known [for a detailed review see3].

Biophysics: Our previous measurements obtained from stationary noise analysis of whole cell recordings suggested a very low single-channel conductance (~ 1 pS) and a high channel density of 50,000 to 100,000 channels/cell [2, 3]. However, stationary noise analysis can seriously underestimate single-channel conductance. In C_6 glioma cells, stationary noise analysis predicted a unitary conductance of ~1 pS [12], whereas both non-stationary noise analysis and direct single-channel measurements in outside-out and whole cell patches showed the existence of 55 pS channels [12]. Recent data indicate that also the endothelial VRAC is a medium conductance channel ($\cong$ 50 pS) as for other tissues. VRAC is further characterised by inactivation at positive potentials and a voltage-dependent recovery from inactivation. Its kinetic properties are modulated by extracellular divalent cations, extracellular pH, the

permeating anion and various channel blockers [13]. We have shown that annexin II, a constituent of the plasma membrane cytoskeleton, is involved in regulation of VRAC [14].
Pharmacology: The commonly used "classical" Cl^- channel blocker, DIDS, has a low affinity for VRAC. Substances, such as NPPB, flufenamic acid, niflumic acid, 1,9-dideoxyforskolin, verapamil and gossypol [2, 15] induce half-maximal block of VRAC at concentrations ranging from a few μM to mM. Inhibitors of PLA_2 such as (pBPB) are efficient blockers of VRAC [2, 4, 16]. The block of VRAC by the anti-estrogen tamoxifen, a compound that has been widely used in the treatment of breast cancer, correlates nicely with its inhibitory effect on the proliferation of endothelial cells [17]. Surprisingly, the anti-malarials quinine and quinidine are also potent blockers of VRAC. Both drugs inhibit the current more efficiently at alkaline extracellular pH, indicating that they exert their action in the uncharged form [18]. Also, the anti-arrhythmic, anti-proliferative and anti-ischemic Ca^{2+}-antagonist mibefradil efficiently inhibits VRAC [19]. Unfortunately, all these compounds lack specificity and a sufficiently high affinity necessary for a biochemical purification of VRAC.
Cellular physiology: The activity of VRAC is an important determinant of the membrane potential both in resting cells and in cells challenged with an hypotonic stimulus [20]. Inhibitors of VRAC have striking, hyperpolarizing effects on the membrane potential of endothelial cells [19, 20]. We described a close correlation between the blocking potency of various Cl^- channel blockers and their inhibitory effect on endothelial cell proliferation. In addition, we could demonstrate that $I_{Cl,swell}$ is drastically down-regulated if muscle cells switch from a proliferating to a differentiated state [2, 3, 18, 21].

MAIN FOCUS OF OUR PRESENT WORK AND PERSPECTIVES

Single channel properties: The underestimation of the single-channel conductance obtained from stationary noise analysis results from the incorrect assumption that the graded increase in whole-cell current is due to a graded increase in open probability of a fixed number of channels. In contrast, an increase in current during cell swelling most likely may originate from an increase in the number of active channels rather than from an increased open probability.
We have already started the re-evaluation of single channel properties which is probably in the range of 50 pS. Alternatively to single channels protocols which exploit cell swelling, we use isovolumic stimulation by intracellular perfusion with GTPγS or reduced ionic strength followed by patch excision at its peak and recording of channel activity in outside-out patches.
Gating mechanism of VRAC: We have been able to show that in the complete absence of intracellular Ca^{2+}, cell swelling cannot activate VRAC. A permissive intracellular concentration of Ca^{2+} (between 50 and 100 nM) is required to activate the channel by volume increase [22]. VRAC can be opened by three different stimuli: cell swelling, a reduction of intracellular ionic strength and GTPγS. The latter two stimuli activate $I_{Cl,swell}$ without any changes in cell volume. In endothelium, thrombin potentiates VRAC via a receptor-dependent mechanism [23]. The activation mechanism is still unclear. We have shown that the protein tyrosine kinase (PTK)

inhibitors tyrphostin B46, tyrphostin A25 and genistein inhibited $I_{Cl,swell}$. Tyrphostin A1, a tyrphostin analogue with little effect on PTK activity, and daidzein, an inactive genistein analogue, were without effect on $I_{Cl,swell}$. The protein tyrosine phosphatase (PTP) inhibitors Na_3VO_4 and dephostatin potentiated $I_{Cl,swell}$ when the current was preactivated by mild hypotonicity. Intracellular perfusion with GTPγS (100 μM) transiently activated a Cl^- current with an identical biophysical and pharmacological profile as $I_{Cl,swell}$. This current was also inhibited by the tested PTK inhibitors and potentiated by the PTP inhibitors. Hypertonicity induced cell shrinking completely inhibited the GTPγS-activated Cl^- current. Intracellular perfusion with GDPβS (1 mM) caused a time-dependent inhibition of $I_{Cl,swell}$, which was more pronounced when the current was activated by mild hypotonicity. These results demonstrate that the activity of endothelial swelling-activated Cl^- channels is dependent on tyrosine phosphorylation and suggest that a G-protein regulates the sensitivity to cell swelling. [24]. It is concluded that cell swelling activates a protein tyrosine kinase and that this might be an essential step in the activation cascade of VRAC.

Under isovolumic conditions $I_{Cl,swell}$ can also be activated by intracellular perfusion with GTPγS [24, 25] or with a pipette solution of reduced ionic strength, Γ_i [26]. Reducing Γ_i at constant osmolarity and Cl^- concentration activates an outwardly rectifying current that is mainly carried by Cl^--ions and inactivates at positive potentials. The permeability ratio for various anions is $P_I > P_{Br} > P_{Cl} >> P_{gluc}$. Blockers of the swelling-activated Cl^- current in CPAE cells also inhibit the current which is activated by a reduction in Γ_i . The protein tyrosine kinase inhibitors tyrphostin B46 and genistein also inhibited the Γ_i induced current. Hypertonic extracellular solutions rapidly and reversibly antagonised the Γ_i activated current, whereas increasing Γ_i precluded activation of $I_{Cl,swell}$ by hypotonic shock. These novel experiments show that a reduction of Γ_i activates an anion current that is identical with that activated by cell swelling. Changes in intracellular ionic strength may shift the volume set point for activation of VRAC. The activation of $I_{Cl,swell}$ by GTPγS and reduced ionic strength requires a tyrosine phosphorylation step since inhibitors of protein tyrosine kinases counteract their effect. These observations (summarised in the table) indicate that tyrosine kinases are required for activation of VRAC.

	VRAC			Tyrosine phosphorylation	
	Activation	**PTK inhibitors**	**C3 exoenzyme**	**p125FAK phosphorylation**	**C3 exoenzyme**
cell swelling	+	Block	Reduction of iodide efflux	+	Reduction
GTPγS	+	Block	ND	+	Block
Ionic strength	+	Block	ND	ND	ND

Effect of cell swelling, GTPγS and reduced ionic strength on $I_{Cl,swell}$ and tyrosine phosphorylation. Modulation by Protein Tyrosine Kinase (PTK) inhibitors, C3 exoenzyme (an irreversible inhibitor of Rho-type GTP-binding proteins). ND: not determined. References, see text.

Flux measurements have provided evidence for the involvement of Rho-type GTP-binding proteins in the activation of $I_{Cl,swell}$ [27]. Pretreatment with Clostridium botulinum C3 exoenzyme (an irreversible inactivator of p21Rho) reduced the swelling-induced efflux of iodide [27]. This suggests that a Rho protein can modulate the tyrosine kinase that is required for activation of $I_{Cl,swell}$. Interestingly, the GTPγS-stimulated tyrosine phosphorylation of $p125^{FAK}$ also requires a functional Rho protein since pretreatment with Clostridium botulinum C3 exoenzyme strongly reduced tyrosine phosphorylation of FAK [28]. In view of the effects of Clostridium botulinum C3 exoenzyme on swelling-induced tyrosine phosphorylation and iodide efflux we will address the role of Rho-type GTP-binding proteins in the activation process of $I_{Cl,swell}$ in endothelial cells. We have already isolated from bovine brain cDNAs for RhoA [29] and RhoGDI, an inhibitor of Rho proteins [30]. As another tool, we will use the cytotoxic necrotizing factor CNF1 which induces deamination of a glutamine at position 63 of RhoA to give a constitutive active Rho protein [31]. Using these tools we will address the role of Rho proteins by activation of VRAC.

Cell Physiology: We focus in our perspective work on the involvement of VRAC in control of endothelial cell proliferation. There is also growing evidence that VRAC may be involved in the regulation of the cell cycle: (a) Blockers of VRAC that are structurally completely different, inhibit proliferation of endothelial cells, T-cells and microglia [17, 32, 33]. (b) VRAC is strongly up-regulated in cervical cancer cells compared to normal cells [34]. (c) VRAC is down-regulated when cells switch from a proliferating to a non-proliferative, differentiated state [for a review see 2, 3, 21]. Experiments on the influence of VRAC on the cell cycle or of special importance because its role in vascularisation of cancer tissue and its possible inhibition.

Pharmacology: One of our future perspectives is to develop more selective tools for modulation of VRAC. At the moment, the most promising blockers for VRAC are tamoxifen and quinine. It has been demonstrated that malaria infection induces a high capacitance pathway, which shares many properties with the VRAC [35, 36]. This pathway might be important for growth and maturation of the plasmodium parasite and might be also important for mediating an efflux of potentially toxic products of the parasite metabolism [37]. It has also been shown for arylaminobenzoates that block of this pathway correlates with the antimalarial effect [38].

Tamoxifen has been show to inhibit endothelial cell proliferation [17]. This effect could be involved in the described tumour-cell death by inhibiting tumour vascularization [39]. We will explore a series of structural analogues of quinine and tamoxifen hoping to find more selective VRAC blockers that can be used eventually as novel therapeutics and for a biochemical purification of that channel. Central in our interest are blockers, which are related to antimalarials. We will screen the antimalarials on the possible blocking action of VRAC including *quinolines* and *quinine derivatives* with alteration of the ternary amine group. Furthermore, we will screen a) novel *arylaminobenzoates*, b) novel *cinchonine* and *cinchonidine* derivatives, c) novel *stilbene derived zwitterionic* compounds.

In relation to the therapeutical use of tamoxifen as an effective tool against breast cancer, we will develop tomoxifen-related proliferation inhibitors (*clomiphene* with the *cis* and *trans* isomers *zuclomiphene* and *enclomiphene*).

Currents hypothesis

Central in our hypothesis is a signal transduction cascade which includes a ionic strength sensitive volume-sensor which may comprise a small G-protein (rho) and tyrosine kinases.
The figure gives four schemes by which we discuss the possible activation cascade. Opening of the channel is dependent on a activation step of tyrosine kinase(s). Tyrosine phosphatase(s) are involved in the reverse reaction. So far, it cannot be excluded that GTPγS itself induces cell swelling by which VRAC is activated. The most likely explanation is that both stimuli, GTPγS and a reduction of intracellular ionic strength, sensitise the volume-dependent PTK activation [24].

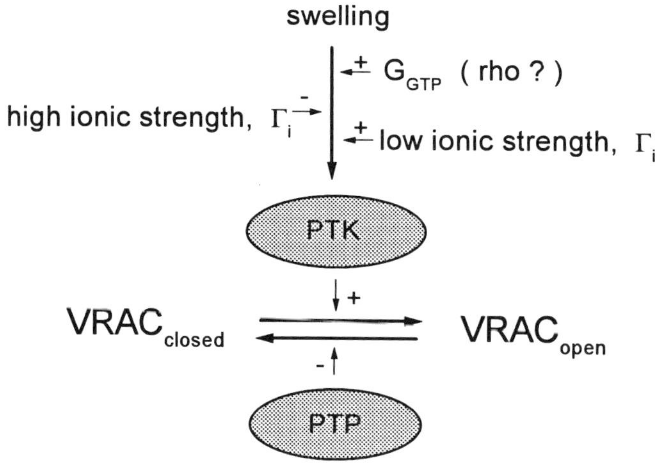

Possible activation cascade of VRAC. PTK: protein tyrosine kinase, PTP: protein tyrosine phosphatase, G_{GTP}: G protein, possibly rho, Γ: ionic strength. See text.

ACKNOWLEDGMENT

We thank Drs D. Trouet, G. Buyse, F. Viana, V. Manolopoulos, M. Oike, and M. Kamouchi for co-operation and many helpful discussions.The excellent experimental contribution of Jean Prenen is essential for the work of this group.

REFERENCES

1 Strange, K, F Emma, and PS Jackson. Cellular and molecular physiology of volume-sensitive anion channels. American Journal of Physiology. American Journal of Physiology 1996, 270:C711-C730.

2 Nilius, B, J Eggermont, T Voets, and G Droogmans. Activation of a Cl^- current by hypotonic volume increase in human endothelial cells. General Pharmacology 1996, 27:1131-1140.

3 Nilius, B, J Eggermont, T Voets, G Buyse, V Manolopoulos, and G Droogmans. Properties of volume-regulated anion channels in mammalian cells. Progress in Biophysics and Molecular Biology 1997, 66:(in the press).

4 Nilius, B, M Oike, I Zahradnik, and G Droogmans. Activation of a Cl^- current by hypotonic volume increase in human endothelial cells. Journal of General Physiology 1994, 103:787-805.

5 De Greef, C, S Vanderheyden, F Viana, J Eggermont, EA Debruijn, L Raeymaekers, G Droogmans, and B Nilius. Lack of correlation between mdr-1 expression and volume- activation of chloride-currents in rat colon cancer cells. Pflügers Archiv European Journal of Physiology 1995, 430:296-298.

6 De Greef, C, J Sehrer, F Viana, K Vanacker, J Eggermont, L Mertens, L Raeymaekers, G Droogmans, and B Nilius. Volume-activated chloride currents are not correlated with P-glycoprotein expression. Biochemical Journal 1995, 307:713-718.

7 Viana, F, K Vanacker, C Degreef, J Eggermont, L Raeymaekers, G Droogmans, and B Nilius. Drug-transport and volume-activated chloride channel functions in human erythroleukemia cells: Relation to expression level of P-glycoprotein. Journal of Membrane Biology 1995, 145:87-98.

8 Buyse, G, C De Greef, L Raeymaekers, G Droogmans, B Nilius, and J Eggermont. The ubiquitously expressed pI_{Cln} protein forms homomeric complexes in vitro. Biochemical and Biophysical Research Communications 1996, 218:822-827.

9 Voets, T, G Buyse, J Tytgat, G Droogmans, J Eggermont, and B Nilius.The chloride current induced by expression of the protein pI_{Cln} in Xenopus oocytes differs from the endogenous volume-sensitive chloride current. Journal of Physiology 1996, 495:441-7.

10 Buyse, G, T Voets, J Tytgat, C De Greef, G Droogmans, B Nilius, and J Eggermont. Expression of human pI_{Cln} and ClC-6 in Xenopus oocytes induces an identical endogenous chloride conductance. Journal of Biological Chemistry 1997, 272:3615-3621.

11 Eggermont, J, G Buyse, T Voets, J Tytgat, H De Smedt, G Droogmans, and B Nilius. Alternative splicing of ClC-6 transcripts generates three truncated isoforms one of which (ClC-6c) is kidney-specific. Biochemical Journal 1997, 325:269-276.

12 Jackson, PS, and K Strange. Single-channel properties of a volume-sensitive anion conductance. Current activation occurs by abrupt switching of closed channels to an open state. Journal of General Physiology 1995, 105:643-660.

13 Voets, T, G Droogmans, and B Nilius. Modulation of voltage-dependent properties of a swelling-activated Cl^- current. Journal of General Physiology 1997, 110:327-336.

14 Nilius, B, V Gerke, J Prenen, G Szücs, S Heinke, K Weber, and G Droogmans. Annexin II modulates volume-activated chloride currents in vascular endothelial cells. Journal of Biological Chemistry 1996, 271:30631-30636.

15 Szücs, G, G Buyse, J Eggermont, G Droogmans, and B Nilius. Characterisation of volume-activated chloride currents in endothelial cells from bovine pulmonary artery. Journal of Membrane Biology 1996, 149:189-197.

16 Oike, M, G Droogmans, and B Nilius. Mechanosensitive Ca^{2+} transients in endothelial cells from human umbilical vein. Proceedings of National Academy of Sciences of the USA 1994, 91:2940-2944.

17 Voets, T, G Szücs, G Droogmans, and B Nilius. Blockers of volume-activated Cl^- currents inhibit endothelial cell proliferation. Pflügers Archiv European Journal of Physiology 1995, 431:132-134.

18 Voets, T, G Droogmans, and B Nilius. Potent block of volume-activated chloride currents in endothelial cells by the uncharged form of quinine and quinidine. British Journal of Pharmacology 1996, 118:1869-1871.

19 Nilius, B, J Prenen, M Kamouchi, F Viana, T Voets, and G Droogmans. Inhibition by mibefradil, a novel calcium channel antagonist, of Ca^{2+}- and volume-activated Cl^- channels in macrovascular endothelial cells. British Journal of Pharmacology 1997, 121:547-555.

20 Voets, T, G Droogmans, and B Nilius. Membrane currents and the resting membrane potential in cultured bovine pulmonary artery endothelial cells. Journal of Physiology 1996, 497:95-107.

21 Voets, T, L Wei, P Desmet, W Vandriessche, J Eggermont, G Droogmans, and B Nilius. Downregulation of volume-activated Cl^- currents during muscle differentiation. American Journal of Physiology 1997, 272:C667-C674.

22 Szucs, G, S Heinke, G Droogmans, and B Nilius. Activation of the volume-sensitive chloride current in vascular endothelial cells requires a permissive intracellular Ca2+ concentration. Pflugers Arch 1996, 431:467-9.

23 Manolopoulos, V, J Prenen, G Droogmans, and B Nilius. Thrombin potentiates volume-activated chloride currents in pulmonary artery endothelial cells. Pflügers Archiv European Journal of Physiology 1997, 433:845-847.

24 Voets, T, V Manolopoulos, J Eggermont, C Ellory, G Droogmans, and B Nilius. Regulation of a swelling-activated Cl-current in bovine endothelium by protein tyrosine phosphorylation and G-proteins. Journal of Physiology Journal of Physiology 1997, 506:341-352.

25 Nilius, B, J Prenen, V T., J Eggermont, KS Bruzik, SB Shears, and G Droogmans.). Ins(1,4,5,6)P4 and Ins(3,4,5,6) inhibt volume-activated chloride currents. American Journal of Physiology 1998, in the press.

26 Nilius, B, J Prenen, T Voets, J Eggermont, and G Droogmans. Reduction of intracellular ionic strength activates the volume-regulated chloride current in cultured bovine pulmonary endothelial cells. Journal of Physiology 1998, 506:353-361.

27 Tilly, BC, MJ Edixhoven, LGJ Tertoolen, N Morii, Y Saitoh, S Narumiya, and HR deJonge. Activation of the osmo-sensitive chloride conductance involves p21(rho) and is accompanied by a transient reorganization of the F-actin cytoskeleton. Mol Biol Cell 1996, 7:1419-1427.

28 Seckl, MJ, N Morii, S Narumiya, and E Rozengurt. Guanosine 5'-3-O-(thio)triphosphate stimulates tyrosine phosphorylation of p125FAK and paxillin in permeabilized Swiss 3T3 cells. Role of p21rho. J Biol Chem 1995, 270:6984-90.

29 Ogorochi, T, Y Nemoto, M Nakajima, E Nakamura, M Fujiwara, and S Narumiya. cDNA cloning of Gb, the substrate for botulinum ADP-ribosyltransferase from bovineadrenal gland and its identification as a rho gene product. Biochemical and Biophysical Research Communications 1989, 163:1175-1181.

30 Fukumoto, Y, K Kaibuchi, Y Hori, H Fujioka, S Araki, T Ueda, A Kikuchi, and Y Takai. Molecular cloning and characterization of a novel type of regulatory protein (GDI) for the rho proteins, ras p21-like small GTP-binding proteins. Oncogene 1990, 5:1321-28.

31 Schmidt, G, P Sehr, M Wilm, J Selzer, M Mann, and k Aktories. Gln 63 of Rho is deamidated by Escherichia coli cytotoxic necrotizing factor-1. Nature 1997, 387:725-29.

32 Schlichter, LC, G Sakellaropoulos, B Ballyk, PS Pennefather, and DJ Phipps. Properties of K^+ and Cl^- channels and their involvement in proliferation of rat microglial cells. Glia 1996, 17:225-236.

33 Schumacher, PA, G Sakellaropoulos, DJ Phipps, and LC Schlichter. Small-conductance chloride channels in human peripheral T lymphocytes. Journal of Membrane Biology. 1995, 145:217-232.

34 Chou, C-Y, S M.-R., and S-N Wu. Volume-sensitive chloride channels associated with human cervical carcinogenesis. Cancer Research 1995, **55**:6077-6083.

35 Kirk, K, HA Horner, BC Elford, JC Ellory, and CI Newbold. Transport of diverse substrates into malaria-infected erythrocytes via a pathway showing functional characteristics of a chloride channel. Journal of Biological Chemistry 1994, 269:3339-3347.

36 Kirk, K, and HA Horner. Novel anion dependence of induced cation transport in malaria-infected erythrocytes. Journal of Biological Chemistry 1995, 270:24270-24275.

37 Ginsburg, H. Transport pathways in the malaria-infected erythrocyte. Biochemical Pharmacology 1994, 48:1847-1856.

38 Kirk, K, and HA Horner. In search of a selective inhibitor of the induced transport of small solutes in Plasmodium falciparum-infected erythrocytes: effects of arylaminobenzoates. Biochemical Journal 1995, 311:761-768.

39 Furman Haran, E, AF Maretzek, I Goldberg, A Horowitz, and H Degani. Tamoxifen enhances cell death in implanted MCF7 breast cancer by inhibiting endothelium growth. Cancer Research 1994, 54:5511-5514.

Cell Volume Regulation: The Molecular Mechanism and Volume Sensing Machinery
Y. Okada, editor.

Molecular Machineries of the Neonatal Rat Kidney against Osmotic Stress

K. Kawahara, N. Anzai, and I. Izumida

Department of Physiology, Kitasato Univ. School of Med.
Kitasato 1-15-1, Sagamihara 228, Japan

INTRODUCTION

Since animal cells are more or less permeable to water, if exposed to hypotonic solutions, they swell due to water-influx, or if exposed to hypertonic solutions, they lose water and shrink. Two types of adaptation to osmotic challenge are postulated: acute adaptation due to activation of ion channels within some seconds (short-term *regulatory volume decrease* (RVD)) and a slower process occurring after several hours of continued exposure to hypertonicity, such as accumulation of amino acids, sugars, polyols, and urea (1-5) (long-term *regulatory volume increase* (RVI)). Cell volume regulation is an intrinsic mechanism to protect the cell in anisosmotic environments. Therefore, most of cells, especially in the kidney medulla, express various types of molecular machineries in the cell membrane, such as ion channels and transporters to regulate cell volume. The major focus of our investigation of the mechanisms for volume regulation in the neonatal rat kidney inner medullary collecting duct (IMCD) cells is on the Ca^{2+}- activated ion channels and both urea and $Na^+/K^+/2Cl^-$ transporters.

ION CHANNELS IN KIDNEY IMCD CELLS

Ion channels play an important role for cell volume regulation, especially for the short-term RVD in various cell types (1-5). A functional relationship between the Ca^{2+}-permeable cation channel and the Ca^{2+}-activated K^+ channel ($K(Ca^{2+})$) has been postulated in choroid plexus (6) and in Necturus kidney proximal tubule cells (7). By using the patch-clamp technique, we identified the single ion channels of both types in the apical cell membrane of the kidney IMCD cells from early neonatal Wistar rats (1-3 days of age)(8). On the contrary, it is interesting that the $K(Ca^{2+})$ channels have not been found in adult rats kidney

IMCD2-3 cells (9). In the excised "inside-out" patches, the non-selective cation channel of 40 pS conducted inward and outward currents of both K^+ and Na^+ but not Cl^- currents. This cation channel was activated by increasing Ca^{2+} on the cytoplasmic side of the patch. The K^+ channel ($K(Ca^{2+})$) of 160 pS was highly selective for K^+ over Na^+ (>20:1). This K^+ channel was also activated by increasing cytosolic Ca^{2+} and depolarizing membrane potentials. The channel activity was inhibited by adding 100 μM Ba^{2+} to the cytoplasmic side. The hypotonic cell-swelling could activate both the non-selective cation channel and the $K(Ca^{2+})$ via increasing intracellular Ca^{2+}, which was followed by an efflux of water. Thus, the swelled cells decrease toward their original cell volume within a few minutes.

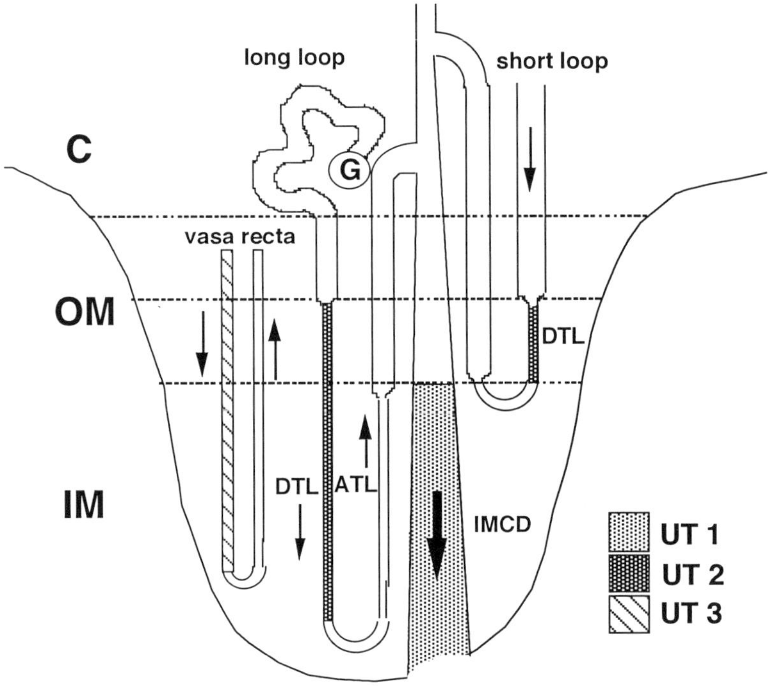

Fig.1. Distribution of three isoforms of urea transporters (UT1, UT2, UT3) in rat kidney medulla. UT1 and UT2 are derived from a same gene by alternative splicing. UT3 is another gene product. Arrows indicate directions of blood and urine flow. Abbreviations, C: cortex, OM: outer medulla, IM: inner medulla, DTL: descending thin limb, ATL: ascending thin limb, IMCD: inner medually collecting duct, G: glomerulus

HYPERTONICITY IN RENAL MEDULLA

Maintenance of a steady-state (isosmotic) condition requires a continuous balancing between the intake of water and the excretion of urine, if evaporation from the body is assumed to be constant. Mammals and birds can survive without any additional water by maintaining their body fluid osmolality through concentration of their urine for a limited period. Therefore, during the period animals are restricted from water, the extrarenal water loss (the body water loss) should be equal to the *free water absorption* by the kidney. As seen in the Eq. (1), the absorption increases roughly in proportion to the product of renal concentrating ability and urine flow.

$$T^{c}H_2O = (U_{osm} - P_{osm})/P_{osm} \times V \qquad (1)$$

,where U_{osm} and P_{osm} are the total osmolality (mosmol/kgH_2O) of urine and plasma; V is volume flow of urine (μl/min) voided in a given collection.
Since the tubules and vasa recta in the kidney run parallel to the long axis of the medulla (papilla) and provide a counter current system (Fig.1), the kidneys can produce the concentrated urine and, consequently, the high osmolalities exist in the kidney medulla (>2000 mosmol/kgH_2O in adult rats)(10). Thus, the tubules and blood vessels including blood cells in this region are exposed to extreme hypertonicity during dehydration.

UREA TRANSPORTERS

In neonatal rats (4-10 days of age), urine osmolality at time 0 (just after deprivation from their mothers) was 225-270 mosmol/kgH_2O which was below the range of serum osmolality of 285-299 mosmol/kgH_2O (Anzai et al., in preparation). This suggests that water balance can be maintained in normal suckling rats without concentrating urine. However, the molecular machineries protecting against osmotic stress, one of the self-defence systems of the cell, should develop in parallel with or in advance of the maturation of renal concentrating ability (10), because urea permeability of the cell may decrease cell deformation under the conditions of anisosmotic stress. By way of comparison, we summarize important findings on urea transporters which, up until now, have been obtained from adult specimens.

Renal distribution of urea transporters mRNA in rat kidney (*adults*): Hediger and his colleagues (11,12) have cloned the urea transporter (UT2) from the rabbit kidney medulla and its splice variant (UT1). They have also isolated the rat homologue (UT3)(13) of HUT11 which was originally cloned from human bone marrow by Martial et al. (14). Vasopressin-regulated urea transporter (UT1) is localized in the IMCD, whereas UT2, which is not induced by AVP, is localized in the descending thin limb (DTL) of Henle's loop (12,15). Phloretin-sensitive urea transporter (UT3) is expressed in vasa recta of the kidney outer (OM) and inner medulla (IM) (13). Shayakul et al. (16) have isolated a novel urea transporter isoform (UT4) which may be localized in the basolateral membrane of the IMCD cells. At least four isoforms of urea transporters are expressed in the kidney.

Expression and renal distribution of urea transporter mRNAs (UT1, UT2, UT3) in *neonatal* rat kidney: Northern analysis of rat kidneys using the probe for UT2/UT1 has demonstrated heavy expression of short (UT2) and long (UT1) mRNA in the IM, whereas UT2 mRNA was only expressed in the OM (17). According to the results obtained from subtype-specific probes for UT1 and UT2 (15), these short (2.8 kb) and long (4.0 kb) length mRNAs should be UT2 and UT1, respectively. No expression was observed in the cortex of the rat kidneys (Fig.1).

As evidence of early development of renal concentrating ability, the short and long signals were found in the IM of the 5, 10, and 15 day-old rats and relatively increased in the elder rats. In the OM of the 5, 10, and 15 day-old rats, only the short signal (UT2) was detected. Renal expression of UT3 mRNA was found in the OM and IM.

Effects of dehydration: Expression of urea transporter mRNAs was studied in the OM and IM from rat kidneys after milk deprivation. The short and long mRNAs from the kidney OM and IM region increased when neonatal rats (5-day-old) were restricted from milk (water) for 16 hours. However, only the long signal (UT1) from the IM region increased in response to the identical dehydration when the rats were 15-day-old. Urine osmolality of the 15-day-old rats was hypertonic (about 500 mosmol/kgH_2O) even at the control conditions. This value was significantly higher than that of the younger neonates (225-270 mosmol/kgH_2O).

IMCD cell culture: Kidney IMCD cells were prepared with a slight modification as described by Stokes et al. (18). There were only long signals (4.0 kb length), and no short signals (2.7 kb length) present in the cultured IMCD cells. This indicates that the UT2/UT1 probe only hybridized with the longer mRNA, i.e. UT1. The UT1 signals increased when the cells were cultured at 500 mosm/kgH_2O (+100 mM NaCl) for 12 hours, but did not in hyperosmotic (+200 mM urea) cells. These results suggest that urea transporters are expressed in the early neonatal rats and are regulated by dehydration. Expression of UT1 mRNA was stimulated by hypertonicity (+100 mM NaCl) without ADH in the cultured IMCD cells. Addition of urea (200 mM) to the normal medium failed to yield in increase in the expression of UT1 mRNA. This indicates that urea was osmotically inactive in the present experiments.

Na/K/2Cl TRANSPORTER: A Na/K/2Cl cotransporter plays an important role for cell volume regulation (RVI) under the conditions of hypertonic cell shrinkage in Ehrlich cells (19). Recently, two isoforms of a Na/K/2Cl cotransporter have beencloned: 1) NKCC1 from shark rectal glands (20) or mBSC2 from mIMCD cells in culture (21); and 2) NKCC2 (22) and rBSC1 (23) from rabbit and rat kidneys, respectively. The apical cotransporters (NKCC2, rBSC1) are responsible for the NaCl absorption, whereas the basolateral ones (NKCC1, mBSC2) are responsible for the Cl^- secretion (24). To elucidate the cell volume regulatory role of the basolateral type of the Na/K/2Cl cotransporters in the kidney IMCD, we isolated the rat homologue (rBSC2) of mBSC2 using a RT-PCR technique. Northern blot analysis showed that rBSC2 mRNA was, as expected from Delpire et al. (21), expressed ubiquitously in rat tissues, including brain, trachea, lung, kidney, stomach, and so on (Anzai, in preparation). When the cultured IMCD cells were exposed to a hypertonic environment (+100 mM NaCl) for 12 hours, rBSC2 mRNA increased by 4-5 times. Urea (200 mM) failed to increase the mRNA amounts in the same experimental conditions. Our results confirm that the Na/K/2Cl cotransporter (basolateral type) plays a role in cell volume regulation (RVI) under chronic hypertonicity, and that urea may be osmotically inactive due to the high urea permeability of the IMCD cells.

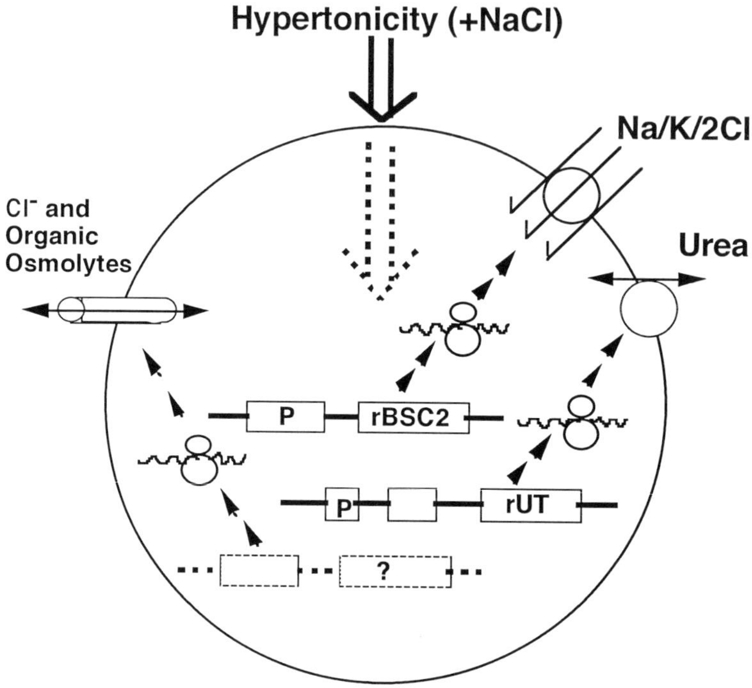

Fig.2. A model of rBSC2 and rUT gene regulation by effective hypertonicity in rat IMCD cells (a slow process ocurring after several hours of continued exposure to hypertonicity). Addition of NaCl (100mM) to culture medium may activate transcription from the rBSC2 and rUT gene. "P" indicates a putative promotor region of each gene.

CONCLUSION AND PERSPECTIVE

Figure 2 illustrates the molecular machineries which play a role in cell volume regulation under the conditions of chronic hypertonicity. Northern hybridization analyses revealed that the rBSC2 and UT1 mRNAs were 4-5 times more abundant in the hypertonic (+NaCl) than isotonic IMCD cells. However, it is not known what the relationship is between the transcription rate of each mRNA and the compatible urea and ion transport rates. Addition of urea (+200 mM) to the culture medium failed to stimulate the transcription rate in either rBSC2 or UT1. This suggests that cell deformation due to hypertonicity in the extracellular environment is an effective signal to begin transcription. There remain unsolved questions regarding how mammalian cells recognize hypertonicity (what are the molecular machineries of the cell membrane for sensing hypertonicity?) and how the signal is transduced within the cell (what is the transcriptional machinery which is stimulated by the hypertonic environment?). Although three independent cascades of MAP kinase homologues may be activated by hypertonicity (25,26), it is not clear at this point whether any of these pathways are involved in activation of a putative promotion region of the rBSC2 and UT genes.

ACKNOWLEDGEMENTS

Work from the authors laboratory in this article was partly supported by CREST from Japan Science and Technology cooperation (JST) and by the Frontier Project of Kitasato University, School of Med. from the Ministry of Education, Culture, and Science to K.K.

REFERENCES

1 Macknight ADC, Gordon LGM, Purves RD. J Experi Zoology 194; 268: 80-89.
2 Kwon H-M, Handler JS. Current Opinion in Cell Biol 1995; 7: 465-471.
3 Hoffmann EK, Dunham PB. Intl Rev Cyto 1995; 161: 173-262.
4 Strange K, Emma F, Jackson PS. Am J Physiol 1996; 270: C711-C730.
5 Okada Y. Am J Physiol 1997; 273: C755-C789.
6 Christensen O. Nature 1987; 330: 66-68.
7 Filipovic D, Sackin H. Am J Physiol 1992; 262: F857-F870.
8 Kawahara K, Onodera M, Aouda A, Fukuda Y. FASEB J 1994; 8: A529.
9 Nonaka T, Matsuzaki K, Kawahara K, Suzuki K, Hoshino M. Biochim Biophys Acta 1995; 1233: 163-174.
10 Edwards BR, Mendel DB, LaRochelle, FTJr, Stern P, Valtin H. In: Spitzer A, ed. The Kidney During Development: Morphology and Function. New York: Masson, 1981; 233-240
11 You G, Smith CP, Kanai Y, Lee W-S, et al. Nature 1993; 365: 844-847.
12 Shayakul C, Steel A, Hediger MA. J Clin Invest 1996; 98: 2580-2587.

13 Tsukaguchi H, Shayakul C, Berger UV, Tokui T, et al. J Clin Invest 1997; 99: 1506-1515.
14 Martial S, Olives B, Abarami L, Couriaud C, et al. Am J Physiol 1996; 271: F1264-F1268.
15 Shayakul C, Knepper MA, Smith CP, DiGiovanni SR, Hediger MA. Am J Physiol 1997; 272: F654-F660.
16 Shayakul C, Tsukaguchi H, Hediger MA. J Am Soc Nephrol 1997; 8: 25A.
17 Izumida I, Anzai N, Kokubo A, Kobayashi Y, et al. J Am Soc Nephrol 1997; 8:18A.
18 Stokes JB, Grupp C, Kinne RKH. Am J Physiol 1987; 253: F251-F262.
19 Geck P, Pfeiffer B. Ann New York Acad Sci 1985; 456: 166-182.
20 Xu J-C, Lytle C, Zhu TT, Payne JA, et al. Proc Natl Acad Sci USA 1994; 91: 2201-2205.
21 Delpire E, Rauchman MI, Beier DR, Hebert SC, Gullans SR. J Biol Chem 1994; 269: 25677-25683.
22 Payne JA, Forbush BIII. Proc Natl Acad Sci USA 1994; 91: 4544-4548.
23 Gamba G, Miyanoshita A, Lombardi M, Lytton J, et al. J Biol Chem 1994; 269: 17713-17722.
24 Kaplan MR, Mount DB, Delpire E, Gamba G, Hebert SC. Annu Rev Physiol 1996; 58: 649-668.
25 Terada Y, Tomita K, Homma MK, Nonoguchi H, et al. Kidney Int 1995; 48: 1801-1809.
26 Raingeaud J, Gupta S, Rogers JS, Dickens M, et al. J Biol Chem 1995; 270: 7420-7426.

Cell Volume Regulation: The Molecular Mechanism and Volume Sensing Machinery
Y. Okada, editor.

Cloning and properties of a hypertonicity-inducible inward rectifier K^+ channel from euryhaline eels

Y. Suzuki[a], N. Nakamura[a], H. Sakuta[a], K. Takano[b], T. Fujita[b], and S. Hirose[a]

[a]Department of Biological Sciences, Tokyo Institute of Technology, 4259 Nagatsuta-cho, Midoriku, Yokohama 226-8501, Japan

[b]Fourth Department of Internal Medicine, University of Tokyo School of Medicine, Bunkyo-ku, Tokyo 112, Japan

SUMMARY

Osmoregulation in teleost fish is achieved largely through active salt transport by specialized cells known as "chloride cells". The chloride cells are of special interest in view of cell volume regulation and osmoregulation since they have an excellent ability of sensing cell volume and an extraordinary power of ion transport to cope with osmotic challenge. In an attempt to clarify the adaptive mechanisms, we carried out differential mRNA display to monitor the molecular alterations associated with adaptation of euryhaline eels to different salinities and identified an inward rectifier K^+ channel (eKir) whose expression is highly dependent on salinity and is confined to the osmoregulatory organs such as the gill, kidney, and intestine. The eKir channel cloned here and shown to be unique in its sequence, tissue distribution, and inducible nature appears to be a novel member of the Kir family that is involved in osmoregulation.

INTRODUCTION

Chloride cells, volume sensing, and osmoregulation

Cell volume regulation is an essential feature of most cells [1–3]. This common ability of cells to volume regulate, however, is expected to be greatly exaggerated in specialized cells that face large changes in osmolarity such as cells in the renal inner medulla and small intestine. In fact, the renal and intestinal epithelial cells are frequently used for studying the mechanism of cell volume regulation. Although, in mammals, large changes in osmolarity are only seen within the kidney and the small intestine, in fishes, especially in euryhaline fishes that can survive both in fresh water and in sea water, there are specialized cells that secrete NaCl in response to a highly variable extracellular environment to maintain homeostasis of the internal milieu. The specialized cell is called "chloride cell" [4]. Chloride cells are large, round or ovoid cells that are rich in mitochondria and Na^+,K^+-ATPase reflecting their extraordinary power of ion transport, and mainly located in the basal region of the secondary lamella of the teleost gill and in the opercular epithelium (Fig. 1).

Osmoregulation in teleost fish is achieved largely through active salt transport by

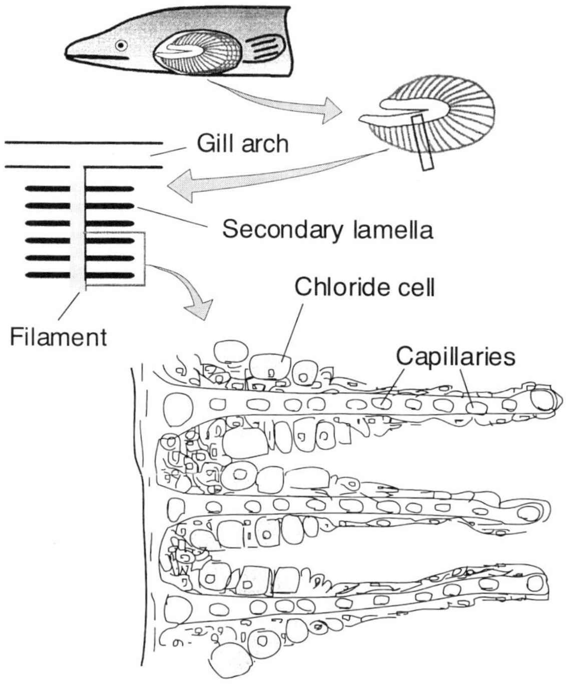

Figure 1. Location and structure of chloride cells. Chloride cells are deeply invaginated by infoldings of the surface membrane from the basal and lateral surfaces and have an extensive tubular system emanating from the infoldings.

chloride cells; for example, salts are actively taken up in fresh water and actively excreted in sea water. To compensate the water loss in salt water environments, sea water-adapted fishes drink sea water and excrete the excess salts from chloride cells in the gill. This causes a transient increase in plasma osmolarity, which in turn is detected on the basolateral side of the chloride cell and results in its shrinkage with subsequent activation of the cell volume regulation system and the salt excretion system. By the analyses using specific inhibitors, the shrinkage of the chloride cell has been demonstrated to activate the Na-K-2Cl cotransporter and Na/H exchanger and require the integrity of apical chloride channels and normal levels of Ca^{2+} [5]. Involvement of a basolateral K^+ channel has also been proposed. The Cl/HCO_3 exchanger, known to be an important element in cellular volume regulation, does not appear to participate in the osmotic response of chloride cells to high salinity [5].

Understanding the molecular basis of these extraordinary abilities of euryhaline species has long been a major goal in animal physiology. Previous studies were mainly focused on the morphological, electrophysiological, and hormonal aspects. Although such conventional

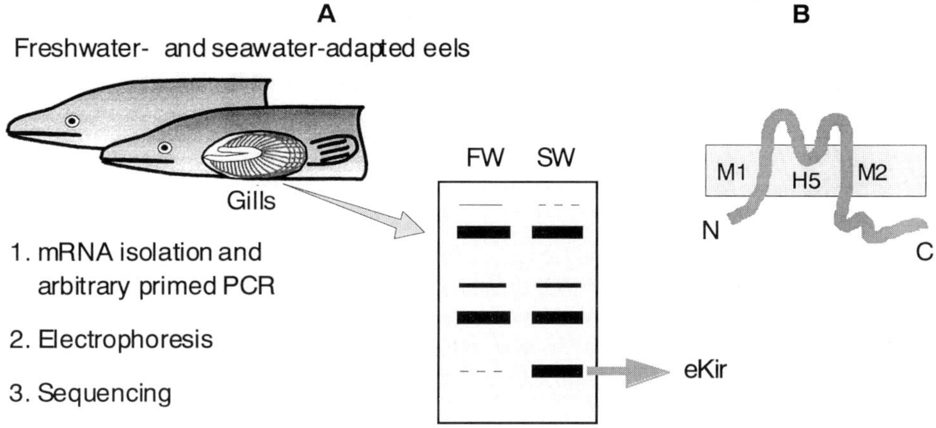

Figure 2. Outline of the technique of differential display (**A**) used to determine differentially expressed mRNA species between comparative populations of cells and tissues. **B**, Membrane topology of inward rectifier K^+ channel.

approaches have revealed a number of interesting facts concerning the osmoregulatory processes, detailed mechanisms and the molecules involved remain to be clarified.

RESULTS AND DISCUSSION

Identification of a K^+ channel highly expressed in seawater-adapted eels

Recently a method called differential display [6, 7] has been developed for identifying altered gene expression at the mRNA level. We applied this methodology for identifying genes that are strongly induced during adaptation of eels to sea water. Fig. 2 shows the outline of the differential display we carried out on mRNA preparations from the gills of eels adapted to fresh water or sea water. The overall patterns between the two mRNA preparations were similar but there were several bands whose intensities were either increased or decreased in sea water eel gills. One of the differentially displayed products which exhibited a particularly strong dependence on the salinity for its expression was isolated and sequenced. A data base search using the BLAST program revealed a weak but significant similarity between the differential display product and certain members of the inward rectifier K^+ channel (Kir) [8, 9]. Identification and characterization of an ion channel that is differentially expressed in fresh water and sea water eels may yield novel molecular clues that may be useful in understanding osmoregulation. We therefore cloned the corresponding full-length cDNA and characterized it.

Properties of hypertonicity-inducible eel K^+ channel

Sequencing of the cloned cDNA and electrophysiological analyses of the protein product expressed in a cell line derived from the GH_3 cell indicated that the clone identified by differential display encodes a 372-amino-acid inward rectifier K^+ channel. We named the eel channel "eKir". Like other members of the Kir family, it has two transmembrane spanning segments (M1 and M2) and a pore region (H5) (Fig. 3). The H5 pore-lining region exhibits the highest degree of sequence conservation among the Kir sequences so far determined; the other regions including the M1 and M2 are much less conserved. eKir has no N-terminal signal sequence and is therefore likely to have a membrane topology with an intracellular N terminus, a feature common to the members of the Kir family (Fig. 2B). Fig. 4 shows electrophysiological characteristics of eKir. The current-voltage (I-V) characteristics of the whole cell current showed a pronounced inward rectification (Fig. 4). The reversal potential of eKir currents measured in varying concentrations of external K^+ followed the equilibrium potential for K^+ (E_K) predicted from the Nernst equation, suggesting a high selectivity for K^+.

Time course of induction and tissue distribution of eKir

When eels are transferred from fresh water to sea water, they show a transient increase in their plasma osmolarity, which reaches a maximum 10–20 h after transfer, subsequently declines, and returns to near basal levels within 2–3 days. The transient osmolarity increase in plasma is due to the fact that eels drink sea water, absorb the salts through the intestine, and eliminate excess NaCl through the gill. The NaCl excretion process is thought to be triggered by the shrinkage of the chloride cells in the gill and opercular membrane. Roughly in parallel with this time course of the recovery in plasma osmolarity, the induction of eKir expression occurred over a time course of hours to days and lasted while eels were kept in sea water.

To begin to elucidate physiological roles of eKir, we next determined its tissue distribution by RNase protection assay using total RNA preparations from the brain, gill, heart, liver, stomach, intestine, head kidney, and kidney. Among the tissues examined, relatively high levels of the eKir message were detected in the gill, kidney, and intestine of sea water eels; the other tissues and the tissues of fresh water eels contained very low or undetectable levels. The inducible nature and the tissue distribution pattern suggest that eKir may play an important role in osmoregulation. Concerning whether eKir serves a direct function in sensing and regulation of cell volume, we have no direct evidence and therefore the question should be addressed by future studies.

Phylogenetic consideration

Phylogenetic analysis of the mammalian Kir family members so far identified has indicated that they fall into 6 subfamilies with differing electrophysiological properties, mechanisms of activation, tissue distributions, and roles in basic cellular processes: the Kir1–Kir6 families. The Kir1 subfamily is composed of ROMK1 and its splice variants that occur in the kidney; the Kir2 subfamily, IRK1–3, classical types in excitable cells; the Kir3 subfamily, GIRK1–4, G protein-coupled type in excitable cells; the Kir4 subfamily, BIR10;

```
 eKir : MSSTAVQYTTVQSS--DRPVTLSV----GMGPGQTD------RKPRYIQRDGSCRLALEHVPGHL  53
 BIR9 : M-SYYGSSYRIVNV--DS-KYP------GYPPEHAIAE-KRRARRLLHKDGSCNVYFKHIFGEW  54
 IRK1 : MGSVRTNRYSIVSSEEDGMKLATMAVANGFGNGKSKVHTRQQCRSRFVKKDGHCNVQFINV-GEK  64
GIRK1 : MSA-LRRKFGDDYQVVTTSSSGSGLQPQGPGQGPQQQLVPKKKRQRFVDKNGRCNVQHGNLGSET  64
ROMK1 : MGA--SER-S-VFRVLIRALTERMFK-HLRRWFITHIFGRSRQRARLVSKEGRCNIEFGNVDAQS  60

                                 M1
 eKir : G-LYVTDIFTTLVEIRWRVMFLLFSLSYILSWLFFGLLFW-IIALAHGDL---VGLSDG-PCVDN  112
 BIR9 : G-SYMVDIFTTLVDTKWRHMFVVFSLSYILSWLIFGSIFW-LIALHHGDL---LSDPDITPCVDN  114
 IRK1 : GQRYLADIFTTCVDIRWRWMLVIFCLAFVLSWLFFGCVFW-LIALLHGDL---DTSKVSKACVSE  125
GIRK1 : S-RYLSDLFTTLVDLKWRWNLFIFILTYTVAWL-FMASMWWVIAYTRGDLNK-AHVGNYTPCVAN  126
ROMK1 : RFIFFVDIWTTVLDLKWRYKMTVFITAFLGSWFLFGLL-WYVVAYVHKDLPEFYPPDNRTPCVEN  124

                H5                    M2
 eKir : VRSFTAAFLFSLETQTTIGYGFRGMTENCAVAIALVTVQDVLSCFIDTVVIGIVVAKMASARKRA  177
 BIR9 : VHSFTAAFLFSLETQTTIGYGYRCVTEECSVAVLTVILQSILSCIINTFIIGAALAKMATARKRA  179
 IRK1 : VNSFTAAFLFSIETQTTIGYGFRCVTDECPIAVFMVVFQSIVGCIIDAFIIGAVMAKMAKPKKRN  190
GIRK1 : VYNFPSAFLFFIETEATIGYGYRYITDKCPEGIILFLFQSILGSIVDAFLIGCMFIKMSQPKKRA  191
ROMK1 : INGMTSAFLFSLETQVTIGYGFRFVTEQCATAIFLLIFQSILGVIINSFMCGAILAKISRPKKRA  189

 eKir : QTVGFSSCAVINRRDGHVCLSWRLGDFRKNHMVEGVATAQIVRLQQQTSG---EVGVVYQDLPL-  238
 BIR9 : QTIRFSYFALIGMRDGKLCLMWRIGDFRPNHVVEGTVRAQLLRYSEDSEG---RMTMAFKDLKL-  240
 IRK1 : ETLVFSHNAVIAMRDGKLCLMWRVGNLRKSHLVEAHVRAQLLKSRITSEGEYIPLDQIDINVGFD  255
GIRK1 : ETLMFSEHAVISMRDGKLTLMFRVGNLRNSHMVSAQIRCKLLKSRQTPEGEFLPLDQLELDVGFS  256
ROMK1 : KTITFSKNAVISKRGGKLCLLIRVANLRKSLLIGSHIYGKLLKTTITPEGETIILDQTNINFVVD  254

 eKir : -QQSDILLATPTTVTHLIVPGSPLYGLSMLDLYREDFELVVTFTYTGDSTGILHQTRSSYRPAQI  302
 BIR9 : -VNDQIILVTPVTIVHEIDHESPLYALDRKAVAKDNFEILVTFIYTGDSTGTSHQSRSSYVPREI  304
 IRK1 : SGIDRIFLVSPITIVHEIDEDSPLYDLSKQDIDNADFEIVVILEGMVEATAMTTQCRSSYLANEI  320
GIRK1 : TGADQLFLVSPLTICHVIDAKSPFYDLSQRSMQTEQFEVVVILEGIVETTGMTCQARTSYTEDEV  321
ROMK1 : AGNENLFFISPLTIYHIIDHNSPFFHMAAETLSQQDFELVVFLDGTVESTSATCQVRTSYVPEEV  319

 eKir : RWGARFKDVL-LVNGRSYRADYTLFHETVPVQLHAR-SAQDQDRCKPDNPATTPALTNGQAPPNA  365
 BIR9 : LWGHRFHDVL-EVKRKYYKVNCLQFEGSVEVY-APFCSAKQLDWKDQQLNNLEKTSPAPRILHLG  367
 IRK1 : LWGHRYEPVL-FEEKHYYKVDYSRFHKTYEVPNTPLCSARDLAEKKYILSNANSFCYENEVALTS  384
GIRK1 : LWGHRFFPVISLEE-GFFKVDYSQFHATFEVP-TPPYSVKEQEEMLLMSSPLIAPAITNSKERHN  384
ROMK1 : LWGYRFVPIVSKTKEGKYRVDFHNFGKTVEVE-TPHCAMCLYNEKDARARMKRGYDNPNFVLSEV  383

 eKir : SPTHATP                                                           372
 BIR9 : HQHQEEILQRSCHGEQL                                                 384
 IRK1 : KEEEEDSENGVPESTSTDSPPGIDLHNQASVPLEPRPLRRESEI                      428
GIRK1 : SVECLDGLDDISTKLPSKLQKITGREDFPKKLLRMSSTTSEKAYSLGDLPMKLQRISSVPGNSEE  449
ROMK1 : DETDDTQM                                                          391

GIRK1 : KLVSKTTKMLSDPMSQSVADLPPKLQKMAGGPTRMEGNLPAKLRKMNSDRFT              501
```

Figure 3. Amino acid sequence comparison of eel Kir (eKir) with those of other members of the inward rectifier K^+ channel family. The GenBank accession numbers are: eKir, AB009669; BIR9, X83581; IRK1, X73052; GIRK1, L25264/U01071; ROMK1, X72341. For the membrane topology, see Fig. 2B.

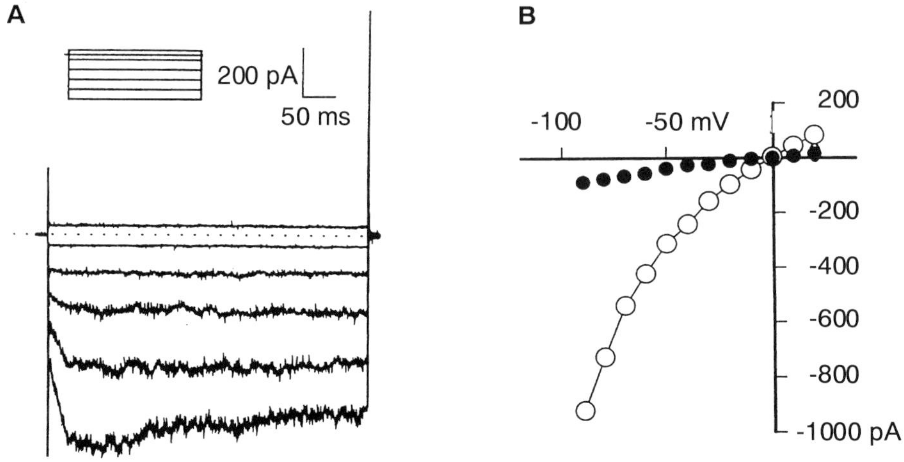

Figure 4. Electrophysiological characterization of eKir. **A**, membrane current from a eKir-transfected cell under voltage clamp in the extracellular solution containing 140 mM K^+. The membrane currents were evoked by pulse steps to +10, -10, -30, -50, -70, -90 mV from the holding potential of 0 mV. Dotted line indicates the zero current level. **B**, I-V relationship of the membrane current from the eKir-transfected cell (open circle). Closed circle shows the I-V relationship of the currents from the vehicle-transfected cells.

the Kir5 subfamily, BIR9, highly expressed in the kidney; and the Kir6 subfamily, uK_{ATP}-1 and Kir6.2 with ATP sensitivities.

Fig. 5 shows a phylogenetic tree of the known members of the Kir family including the newly cloned eKir. The position of the eKir in the tree is far from those of the other members (50% similarity). It seems reasonable, therefore, to assume that eKir is a previously unreported member of the Kir family and belongs to a new subfamily (Kir7) that diverged early during evolution from the Kir5 subfamily. However the possibility remains that, although the sequence divergence of eKir is very high, it may simply reflect the species difference since no eel Kir sequences other than eKir were determined. We therefore cloned, by polymerase chain reaction (PCR), eel homologs of the members of the Kir family and compared their sequences with the mammalian counterparts. Phylogenetic analysis of the PCR products demonstrated that despite the evolutionary distance between the eel and mammals, their members of the Kir subfamilies share a high degree of similarity (70–90%, Fig. 5), supporting our assumption that eKir represents a new subfamily of the Kir family. We are currently searching for mammalian homologs of eKir.

Physiological significance (any relation to volume regulation?)

Determination of localization of eKir at the cellular level may provide clues for its physiological roles. Using the salt-adapted eel tissues that were found to contain relatively

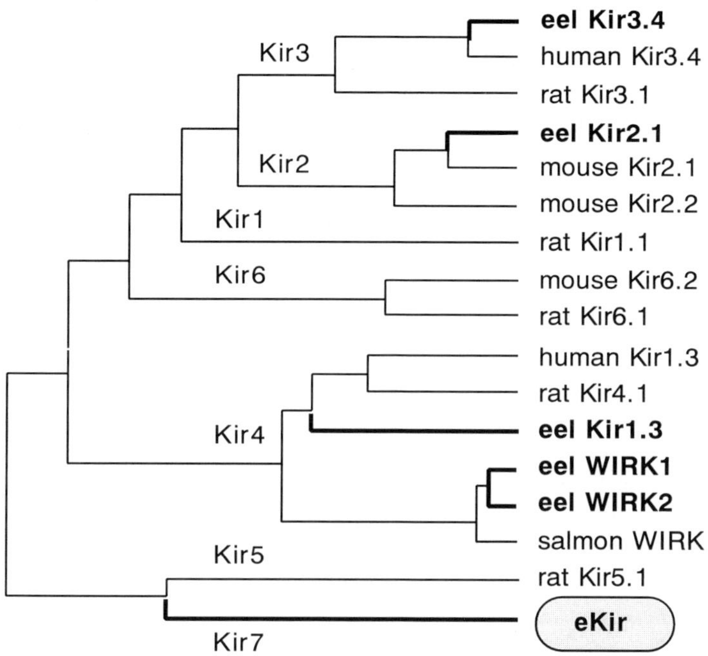

Figure 5. Evolutionary tree showing the probable relationships of various members of the Kir family, based on computerized comparisons of their amino acid sequences.

high levels of the eKir message, we performed immunohistochemical staining with an antiserum against recombinant eKir. Preliminary results indicated that in the gill, chloride cells are the most likely site of location. In the kidney and intestine, epithelial cells were densely stained. These locations are the major sites involved in osmoregulation and are known to exhibit cellular polarity. Useful information will be obtained by determination of the apical *versus* basolateral distribution of eKir in these polarized epithelial cells. At present, there is no direct evidence for the involvement of eKir in the cell volume regulation, but there is a possibility that the channel protein is one of the major components of the volume-sensing and regulatory system in specialized cells.

Our identification of an inducible channel protein, eKir, on which euryhaline fishes are supposed to depend in the seawater milieu may serve as a molecular clue for clarifying the molecular basis for seawater adaptation. Future studies should reveal how hypertonicity is recognized and how the recognition is converted to the activation of the eKir gene as well as the physiological functions of the channel in association with other channels and transporters.

Acknowledgments

This work was supported by a grant-in-aid for specially promoted research from the Ministry of Education, Science, Sports and Culture of Japan.

REFERENCES

1 Burg MB. Molecular basis of osmotic regulation. Am J Physiol 1995; 268: F983–F996.

2 Strange K, Emma F, Jackson, PS. Cellular and molecular physiology of volume-sensitive anion channels. Am J Physiol 1996; 270: C711–C730.

3 Okada Y. Volume expansion-sensing outward-rectifier Cl^- channel: fresh start to the molecular identity and volume sensor. Am J Physiol 1997; 273: C755–C789.

4 Perry SF. The chloride cell: structure and function in the gills of freshwater fishes. Annu Rev Physiol 1997; 59: 325–347.

5 Zadunaisky JA. Chloride cells and osmoregulation. Kidney Int 1996; 49: 1563–1567.

6 Welsh J, Chada K, Dalal SS, Cheng R, Ralph D, McClelland M. Arbitrarily primed PCR fingerprinting of RNA. Nucleic Acids Res 1992; 20: 4965–4970.

7 McClelland M, Mathieu Daude F, Welsh J. RNA fingerprinting and differential display using arbitrarily primed PCR. Trends Genet 1995; 11: 242–246.

8 Doupnik CA, Davidson N, Lester HA. The inward rectifier potassium channel family. Curr Opin Neurobiol 1995; 5: 268–277.

9 Jan LY, Jan YN. Cloned potassium channels from eukaryotes and prokaryotes. Annu Rev Neurosci 1997; 20: 91–123.

1998 Elsevier Science B.V.
Cell Volume Regulation: The Molecular Mechanism and Volume Sensing Machinery
Y. Okada, editor.

The Na^+/H^+ Exchanger and Volume Regulation

Shigeo Wakabayashi, Toshitaro Ikeda, Yoko Kobayashi, Takahiro Iwamoto, Munekazu Shigekawa

Department of Molecular Physiology, National Cardiovascular Research Institute, Osaka 565, Japan

The amiloride-sensitive Na^+/H^+ exchanger, which carries out an electroneutral exchange of Na^+ for H^+ across the plasma membrane, play an important role for regulation of intracellular pH (pH_i) and cell volume in virtually all cells (for review, see refs. 1 and 2). A ubiquitous isoform NHE1 is known to be rapidly activated in response to a variety of extracellular stimuli ranging from growth factors to mechanical stimuli such as hyperosmotic stress and cell spreading. The hyperosmotic stress causes initial cell shrinkage followed by gradual volume recovery (Fig. 1). It is believed that the volume recovery is attributable to net influxes of ions such as Na^+ and Cl^- and obligatory movement of water. Thus the Na^+/H^+ exchanger, together with Cl^-/HCO_3^- exchanger, is thought to be an important Na^+/Cl^--entry pathway. The activation of NHE1 in response to osmotic stress has been shown to result from an alkaline shift in pH_i sensitivity of NHE1. The shift in pH_i threshold (set point) is usually explained by increased affinity of the exchanger for the intracellular H^+ at the allosteric "modifier site" that is thought to be different from Na^+ and H^+ transport sites. As this kinetic feature is similar to that for the growth factor-induced activation of NHE1, a mechanism common to both may regulate NHE1 at the level of the exchanger molecule.

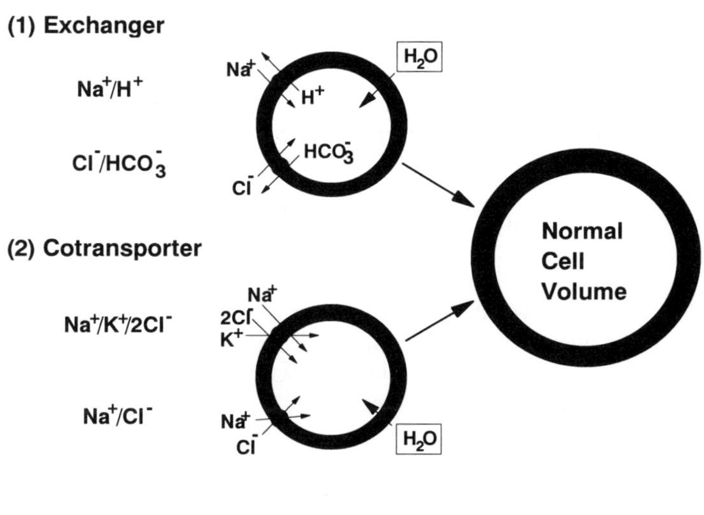

Figure 1. Regulatory volume increase

We have presented evidence that NHE1 is regulated via multiple mechanisms involving distinct cytoplasmic segments of NHE1 (Fig.2) (3-9). (i) Cytoplasmic domain of NHE1 contains two calmodulin (CaM)-binding regions (5). High affinity CaM-binding domain (subdomain III, aa 635-656) is thought to function as an autoinhibitory domain which is removed from the H^+ modifier site in response to an increase in the intracellular Ca^{2+} concentration ($[Ca^{2+}]_i$) (6). This segment could also be involved in the osmotic regulation, since deletion of Ca^{2+}/CaM-binding domain of NHE1 greatly (80%) reduces hyperosmolarity-induced cytoplasmic alkalinization. (5). (ii) Subdomain I is required for phosphorylation-dependent regulation of NHE1, although there is no phosphorylation sites within this subdomain (3,4). We proposed that a phosphorylatable regulatory factor(s) binds to this domain and regulates NHE1. Recently, Lin et al. (10) identified a candidate protein (calcineurin homologue protein, CHP) that binds to this region of NHE1. They suggested that CHP is an inhibitor that is released from NHE1 in response to growth factors. However, it is not clear whether this protein is indeed involved in NHE1 regulation or, if so, how this protein regulates NHE1. (iii) Subdomain IV contains several serine residues that can be phosphorylated in vivo (4). However, phosphorylation at these sites is not required for growth factor- or osmotic stress-induced activation of NHE1.

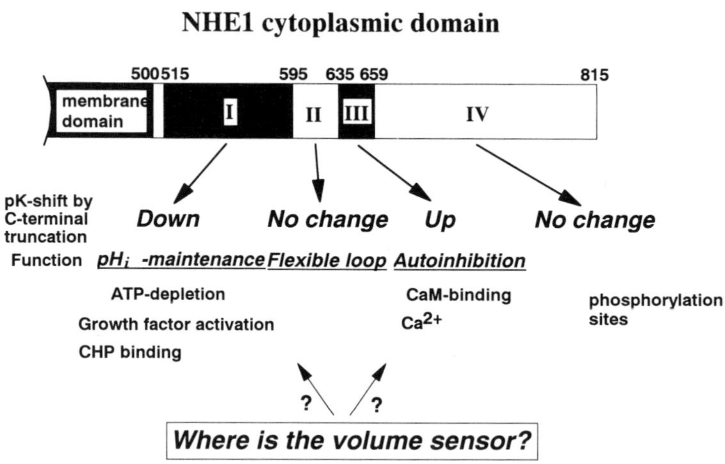

Figure 2. Schematic representation of cytoplasmic subdomains of NHE1.

Recently we have examined whether intracellular Ca^{2+} is involved in the shrinkage (hyperosmolarity)-induced activation of NHE1. Although hyperosmolarity does not increase average $[Ca^{2+}]_i$ monitored by fura-2 fluorescence in CCL39 fibroblastic cells, we cannot rule out the possibility that it increases subplasmalemmal $[Ca^{2+}]$. As a maneuver to reduce the subplasmalemmal $[Ca^{2+}]$, we overexpressed a potent Ca^{2+}-extruder, the Na^+/Ca^{2+} exchanger in CCL39 fibroblasts (11). Expression of NCX1 markedly inhibited a transient rise in average $[Ca^{2+}]_i$ induced by various agents such as α-thrombin. Confocal fluorescence Ca^{2+} imaging revealed that an increase in $[Ca^{2+}]$ in the peripheral cytoplasm was suppressed in NCX1-overexpressing CCL39 cells stimulated with α-thrombin (T. Iwamoto et al. submitted). In these cells, a long-lasting cytoplasmic alkalinization (60-80% inhibition) induced by α-thrombin was markedly inhibited, indicating that a rise in subplasmalemmal $[Ca^{2+}]$ plays a crucial role in thrombin-induced sustained activation of NHE1 (T. Ikeda et al. Am. J. Physiol. submitted). In sharp contrast to the response to thrombin, expression of NCX1 did not inhibit hyperosmolarity-induced alkalinization (Fig. 3). In addition, down regulation of protein kinase C (PKC) did not inhibit hyperosmolarity-induced alkalinization in NCX1 transfected and control CCL39 cells, although the same treatment inhibited thrombin-induced alkalinization partially in control cells and abolished it completely in NCX1-transfected cells. Therefore, the hyperosmolarity-induced activation of NHE1 is totally independent of Ca^{2+} and PKC in CCL39 fibroblasts, whereas the thrombin response is mediated exclusively via Ca^{2+} and PKC. These findings are consistent with observations obtained with other cell types (12,13).

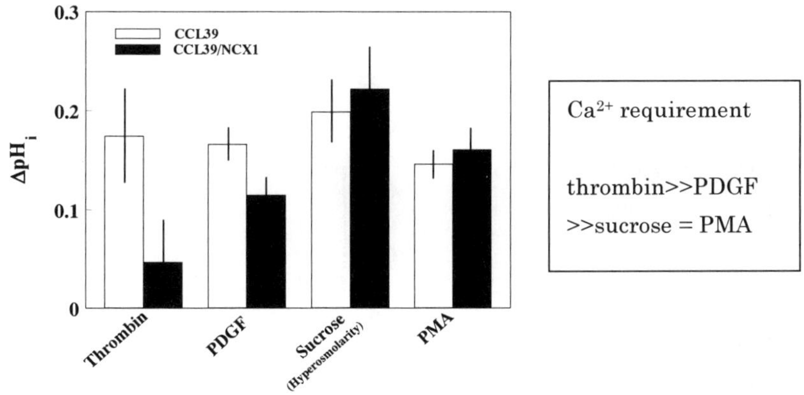

Figure 3. Effect of NCX1-overexpression on cytoplasmic alkalinization induced by various stimuli. Change in pH_i was measured with ^{14}C-benzoic acid.

As shown in Fig. 2, two subdomains (domains I and III) play an important role in determining pH_i-sensitivity of NHE1. Therefore, it may be inferred that one or both of these domains functions as a volume-sensitive region. As described above, in fact, deletion of the CaM-binding domain (domain III) markedly (80%) reduces the extent of osmotic stress-induced alkalinization (5). In addition, osmotic stress failed to induce alkalinization in the Δ595 (truncation of C-terminal region aa 596-815) transfectant, whereas it induced alkalinization normally in the Δ698 (truncation of C-terminal region aa 699-815) transfectant. Therefore, the CaM-binding domain appears to be essential for cell shrinkage-induced activation of NHE1. However, it is difficult to conclude that the CaM-binding domain is the volume-sensing region, because deletion of this domain abolishes its autoinhibitory function, rendering NHE1 constitutive active. To overcome this problem, we constructed chimeric molecules among NHE isoforms. We observed that hyperosmolarity does not induce cytoplasmic alkalinization in cells expressing epithelial isoform NHE2, whereas it induces strong acidification in cells expressing another epithelial isoform NHE3. Hyperosmolarity-induced inhibition of NHE2 and NHE3 was also reported by other groups (14,15). These findings indicate that there exists a NHE1-specific volume-sensitive region within the NHE1 molecule. We used chimeras constructed between NHE1 and NHE2 because sequence homology is considerably high between these two isoforms despite the different response to hyperosmolarity. We replaced domains I and III of NHE1 with homologous regions of NHE2 (Fig. 4). The pH_i-dependence of Na^+/H^+ exchange was similar among all these NHE variants (data not shown) suggesting that these NHE2 segments does not affect the pH_i-sensitivity of NHE1. However, replacement of domain III of NHE1 with the NHE2-segment [N1(1/2) and N1(2/2)] resulted in marked reduction in the hyperosmolarity-induced cytoplasmic alkalinization (Fig. 4). In contrast, similar replacement of domain I [N1(2/1)] did not affect it as long as domain III was not replaced. These findings suggest that CaM-binding domain (domain III) plays a crucial role in hyperosmolarity-induced activation of NHE1, although we do not rule out the possibility that other parts of NHE1 are also involved (16).

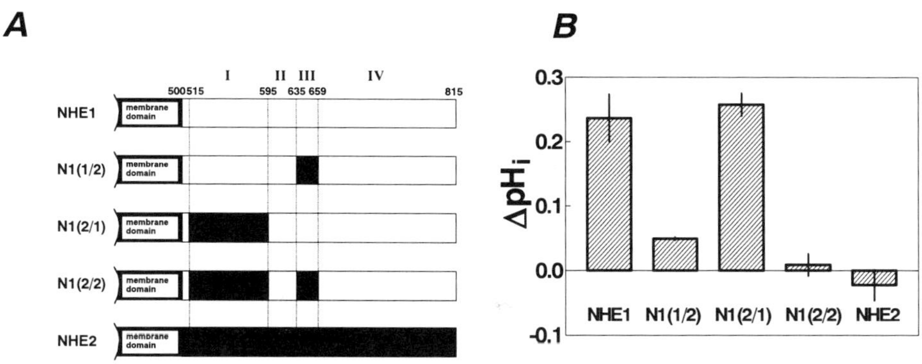

Figure 4. Effect of incorporation of the NHE2-segments on hyperosmolarity-induced cytoplasmic alkalinization. A, schematic representation of chimeric constructs between NHE1 and NHE2. B, hyperosmolarity (200 mM sucrose)-induced change in pH_i was measured by equilibration of ^{14}C-benzoic acid.

As described above, Ca^{2+} is not involved in the osmotic activation. This finding suggests that CaM-binding itself is not involved in the osmotic activation. It is possible, however, that other molecule binds to the CaM-binding domain of NHE1 in response to osmotic stress and thus activates NHE1 via removal of its autoinhibition. Recently, α-actinin has been shown to bind to NMDA-receptor in competition with CaM (17). Such a dual regulation by CaM and a cytoskeletal protein may also occur in NHE1.

Signaling pathway for shrinkage-induced activation of NHE1 is now largely unknown.

Hyperosmolarity-induced alkalinization was not inhibited by G protein (Gi, Go) inhibitor IAP, mitogen/extracellular signaling kinase (MEK1/2) inhibitor PD98059 or phosphatidylinositol 3-kinase inhibitor LY294002. In contrast, it was markedly inhibited by a relatively low concentration (10 - 50 μM) of genistein in CCL39 fibroblastic cells (our data) or human neutrophils (18). Therefore, tyrosine kinases may be involved in hyperosmolarity-induced activation of NHE1, although the NHE1 is not phosphorylated at tyrosine residues. Obviously, further experiments are required to understand the molecular mechanism of shrinkage-induced activation of NHE1, in particular its signaling pathway and the location of built-in "volume-sensor".

REFERENCES

1. Grinstein S, Rotin D, Mason MJ. Biochem. Biophys. Acta 1989; 988: 73-97.
2. Wakabayashi S, Shigekawa M, Pouysségur J. Physiol. Rev. 1997; 77: 51-74.
3. Wakabayashi S, Fafournoux P, Sardet C, Pouysségur J. Proc. Natl. Acad. Sci. USA 1992; 81: 4833-4837.
4. Wakabayashi S, Bertrand B, Shigekawa M, Fafournoux P, Pouysségur J. J. Biol. Chem. 1994; 269: 5583-5588.
5. Bertrand B, Wakabayashi S, Ikeda T, Pouysségur J, Shigekawa M. J. Biol. Chem. 1994; 269: 13703-13709.
6. Wakabayashi S, Bertrand B, Ikeda T, Pouysségur J, Shigekawa M. J. Biol. Chem. 1994; 269:13710-13715.
7. Wakabayashi S, Ikeda T, Noél J, Schmitt B, Orlowski J, Pouysségur J, Shigekawa M. J. Biol. Chem. 1995; 270: 26460-26465.
8. Ikeda T, Schmitt B, Pouysségur J, Wakabayashi S, Shigekawa M,. J. Biochem. 1997; 121: 295-303.
9. Wakabayashi S, Ikeda T, Iwamoto T, Pouysségur J, Shigekawa M. Biochemistry 1997; 36 (42): 12854-12861.
10. Lin X, Barber DL. Proc. Natl. Acad. Sci. USA 1996; 93: 12631-12636.
11. Iwamoto T, Pan Y, Wakabayashi S, Imagawa T, Yamanaka HL, Shigekawa M. J. Biol. Chem. 1996; 271: 13609-13615.
12. Dascalu A, Nevo A, Korenstein R. J. Physiol. Lond. 456: 503-518, 1992.
13. Shrode LD, Klein JD, O'Neill WC, Putnam RW. Am. J. Physiol. 269 (Cell Physiol. 38): C257-C266, 1995.
14. Kapus A, Grinstein S, Wasan S, Kandasamy R, Orlowski J. J. Biol. Chem. 1994; 269: 23544-23552.
15. Nath SK, Hang CY, Levine SA, Yun CHC, Montrose MH, Donowitz M, Tsé CM. Am. J. Physiol. 1996; 270: G431-G441.
16. Bianchini L, Kapus A, Lukacs G, Wasan S, Wakabayashi S, Pouysségur J, Yu FH, Orlowski J, Grinstein S. Am. J. Physiol. 1995; 269 (Cell Physiol. 38): C998-C1007.
17. Wyszynski M, Lin J, Rao A, Nigh E, Beggs AH, Craig AM, Sheng M. Nature 1996; 385: 439-442.
18. Krump E, Nikitas K, Grinstein S. J. Biol. Chem. 1997; 272: 17303-17311.

Part IV
Organic OsmolyteTransporter and Water Channel Involved inVolume Regulation

Cell Volume Regulation: The Molecular Mechanism and Volume Sensing Machinery
Y. Okada, editor.

Transcriptional regulation of transporters for compatible osmolytes by hypertonicity

H. Moo Kwon

Division of Nephrology, Johns Hopkins School of Medicine, 963 Ross Building, 720 Rutland Avenue, Baltimore, Maryland 21205, USA

SUMMARY

When exposed to hypertonicity, renal cells lower the intracellular concentration of electrolytes (or ionic strength) toward isotonic level by accumulating compatible osmolytes including betaine and myo-inositol. Since high cellular ionic strength is deleterious to cell survival, maintaining the cellular ionic strength near the isotonic level is critical in adaptation to hypertonicity. Accumulation of betaine and myo-inositol in renal cells in situ and in culture is mediated by the Na- and Cl-coupled betaine/GABA transporter (BGT1) and Na/myo-inositol cotransporter (SMIT), respectively. Osmo-protective accumulation of betaine and myo-inositol under hypertonic conditions is driven by the increased transcription of the BGT1 and SMIT genes which leads to the increased activity of the transporters. We have studied mechanisms underlying the hypertonicity-induced stimulation of transcription. A tonicity-responsive enahncer (TonE) is identified in the promoter region of the BGT1 gene. Activation of a transcription factor named TonE binding protein (TonEBP) in response to hypertonicity is a key event in the transcriptional regulation: the activated TonEBP binds the TonE site of the BGT1 resulting in stimulation of transcription. Multiple TonEs appear to be involved in the SMIT regulation as well, indicating that TonEBP is the central regulator in adaptation of renal cells to hypertonicity.

CELLULAR MODEL

Figure 1 illustrates the role of compatible osmolytes in adaptation to hypertonicity and cell volume regulation (for review, see 1). When renal cells are moved from an isotonic solution (indicated by 140 mM NaCl) to a hypertonic solution (280 mM NaCl), cells shrink immediately due to osmosis. In many types of cells, cell shrinkage activates transport systems leading to accumulation of K and consequent increase in cell volume via osmosis, a process called regulatory volume increase which is completed within 30 minutes. Regardless of the regulatory volume increase, an important consequence of exposure to hypertonicity is increase in concentration of cell electrolytes or ionic strength (represented by 280 mM K). When the cells are kept in the hypertonic solution for more than several hours, cells slowly accumulate compatible osmolytes. Within cells, increase in the concentration of compatible osmolytes lowers concentration of electrolytes back to the isotonic level. Cellular accumulation of compatible osmolytes betaine and myo-inositol is catalyzed by sodium coupled transporters, BGT1 and SMIT, respectively. Activity of these transporters are markedly stimulated over a period of several hours when renal cells are exposed to hypertonicity, leading to the accumulation of the compatible osmolytes. When these transporters are inhibited, cellular ionic strength remains high leading to cell death. Thus, hypertonicity-induced stimulation of BGT1 and SMIT is very important in protection against the stress of hypertonicity.

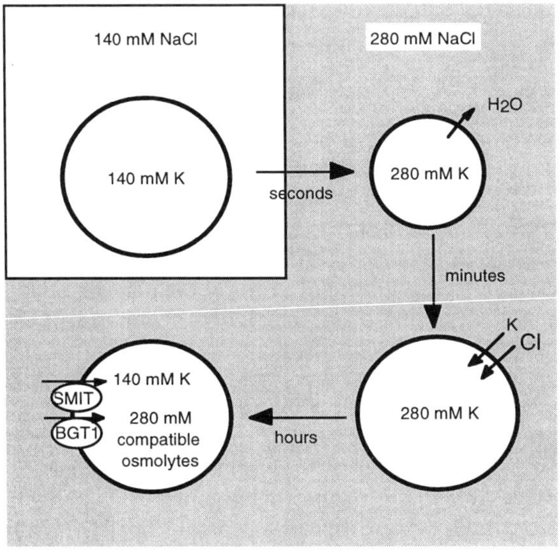

Figure 1: Role of compatible osmolyte transporters, SMIT and BGT1, in adaptation to hypertonicity.

TRANSCRIPTIONAL REGULATION OF BGT1 AND SMIT

To understand molecular mechanisms underlying the osmo-protective accumulation of betaine and myo-inositol, we cloned the coupled transporters using Xenopus oocytes expression (2, 3). BGT1 is a single polypeptides of 614 amino acids and is believed to couple 2 Na and 1 Cl ions for every betaine transported. SMIT, a polypeptide of 718 amino acids, couples two Na ions for every myo-inositol transported (4). When kidney cells are exposed to hypertonicity, transcription of the BGT1 and SMIT genes are markedly stimulated, leading to an increase in mRNA abundance and transport activity (5, 6). Thus, hypertonicity-induced stimulation of transcription is a critical step in the osmo-protective accumulation of compatible osmolytes.

We analyzed the promoter region of the BGT1 gene (7) in order to understand genomic mechanisms involved in the stimulation of transcription. We identified a critical region which contains a novel enhancer that is stimulated by hypertonicity. This 11 base pair enhancer is named tonicity-responsive enhancer (TonE) (8, 9). Based on studies using many mutant TonE sequences, the consensus sequence of TonE was deduced: (C/T)GGAAnnn(C/T)n(C/T). Our recent preliminary data indicate that the SMIT gene is also regulated by several TonE sequences located in the 5' flanking region (10). Other investigators found that TonE sequences regulate the aldose reductase gene (11). AR catalyzes synthesis of sorbitol, a compatible osmolyte. Thus, several genes whose transcription is stimulated by hypertonicity share the same enhancer. Proteins coded by these genes catalyze the accumulation of compatibles.

Since enhancers work by binding to specific transcription factors, we searched for nuclear proteins that bind TonE. Based on a correlation between enhancer activity and binding affinity in a panel of mutant TonE sequences, we identified a couple of slowly migrating bands to be transcription factor(s) named TonE binding protein (TonEBP) (8, 9). Activity of TonEBP is stimulated in parallel with transcription of the BGT1 and SMIT genes when renal cells were exposed to hypertonicity. In vivo methylation assays revealed that the TonE site of the BGT1 is protected from methylation with a time course like that of transcription and TonEBP activity (9). These results establish that hypertonicity activate TonEBP which then bind the TonE site and activate transcription of the BGT1 gene (figure 2). The SMIT and AR genes are likely to be regulated by the same mechanism.

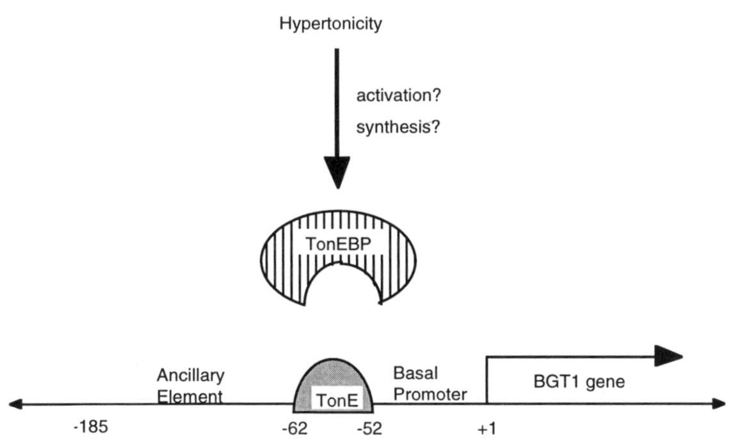

Figure 2: A model of BGT1 regulation by hypertonicity.

HYPERTONICITY SIGNALING PATHWAY

Although a remarkable progress has been made in understanding the hypertonicity-induced stimulation of transcription (figure 2), little is known how tonicity is sensed and how that information is conveyed to TonEBP. More is known in yeast. When cultured in hypertonic media, transcription of *GPD1* (coding for the key enzyme in glycerol synthesis) is stimulated leading to accumulation of glycerol, the major compatible osmolyte in yeast. Genes coding for sensing and signaling of hypertonicity have been identified based on genetic selection (12). Yeast have two

tonicity sensors in the plasma membrane. Of interest, signals from both sensors converge on Hog1, a homolog of MAP kinase.
When renal or non-renal cells are exposed to hypertonicity, three independent MAP kinase homologs - ERK, JNK, and p38 - are activated. But, preventing activation of these MAP kinases does not lead to an inhibition of the hypertonicity-induced transcriptional stimulation of BGT1, SMIT, and AR (13, 14), indicating that, unlike yeast, the MAP kinases are not involved in hypertonicity signaling to transcription in mammals. Work by other investigators suggest that the elevated cellular ionic strength is the signal for the stimulation of transcription for the genes coding for BGT1, SMIT and AR (15). Future work will concentrate on delineating the sensors and signaling pathways linking the increased cellular ionic strength to activation of TonEBP.

ACKNOWLEDGMENT

Work presented in this paper was supported by NIH grants DK44484 and DK42479.

REFERENCES

1. Kwon HM and Handler JS. 1995. Cell volume regulated transporters of compatible osmolytes. Curr. Opinion Cell Biol. 7:465-471.
2. Yamauch A, Uchida S, Kwon HM, Preston AS, Robey R, Garcia-Perez A, Burg M, and Handler JS. 1992 Cloning of a Na- and Cl-dependent betaine transporter that is regulated by hypertonicity. J. Biol. Chem. 267: 649-652.
3. Kwon HM, Yamauchi A, Uchida S, Preston AS, Garcia-Perez A, Burg M, Handler JS. 1992 Cloning of the cDNA for a Na/myo-inositol cotransporter, a hypertonicity stress protein. J. Biol. Chem. 267:6297-6301.
4. Hager K, Hazama A, Kwon HM, Loo D, Handler JS, and Wright E. 1995. Kinetics and specificity of the renal Na/myo-inositol cotransporter expressed in Xenopus oocytes. J. Memb. Biol. 143: 103-113.
5. Uchida S, Yamauchi A, Preston AS, Kwon HM, and Handler JS. 1993. Medium tonicity regulates expression of the Na- and Cl-dependent betaine transporter in Madin-Darby canine kidney cells by increasing transcription of the transporter gene. J. Clin. Invest. 91:1604-1607.
6. Yamauch A, Uchida S, Preston AS, Kwon HM, and Handler JS. 1993 Hypertonicity stimulates transcription of gene for Na/myo-inositol cotransporter in MDCK cells. Am. J. Physiol. 264: F20-F23.
7. Takenaka M, Bagnasco S, Preston AS, Uchida S, Yamauchi A, Kwon HM, and Handler JS. 1995. The canine betaine-GABA transporter gene. Diverse mRNA isoforms are regulated by hypertonicity and are expressed in a tissue specific manner. Proc. Natl. Acac. Sci. USA 92:1072-1076.
8. Takenaka M, Preston AS, Kwon HM, and Handler JS. 1994. The tonicity-sensitive element that mediates increased transcription of the betaine transporter gene in response to hypertonic stress. J. Biol. Chem. 269: 29379-29381.

9. Miyakawa H, Woo SK, Chen CP, Dahl SC, Handler JS, and Kwon HM. Cis- and trans-acting factors regulating transcription of the BGT1 gene in response to hypertonicity. Am. J. Physiol. (in press).
10. Atta MG, Rim JS, Lorica V, Mallee JJ, Berry GT, Handler JS, and Kwon HM. 1997. Cloning of the human Na/myo-inositol cotransporter gene and identification of regions regulating transcription in response to hypertonicity. J. Am. Soc. Nephrol. 8:43A (abstract).
11. Ko BCB, Ruepp B, Bohren KM, Gabbay KH, and Chung SSM. 1997. Identification and characterization of multiple osmotic response sequences in the human aldose reductase gene. J. Biol. Chem. 272:16431-16437.
12. Maeda T, Takekawa M, and Saito H. 1995. Activation of yeast PBS2 MAPKK by MAPKKKs or by binding of an SH3-containing osmosensor. Science 269:554-558.
13. Kwon HM, Itoh T, Rim JS, and Handler JS. 1995. The MAP kinase cascade is not essential for transcriptional stimulation of osmolyte cotransporter genes. Biochem. Biophys. Res. Com. 213: 975-979.
14. Kultz D, Garcia-Perez A, Ferraris JD, and Burg MB. 1997. Distinct regulation of osmoprotective genes in yeast and mammals. J. Biol. Chem. 272:13165-13170.
15. Burg MB. 1996. Coordinate regulation of organic osmolytes in renal cells. Kid. Int. 49:1684-1685.

Cell Volume Regulation: The Molecular Mechanism and Volume Sensing Machinery
Y. Okada, editor.

Urinary concentrating defect and organic osmolytes

Takeshi Nakanishi[a], Atsushi Yamauchi[b], Shigeo Yamamoto[b], and Yoshihiro Takamitsu[a]

[a] Department of Kidney and Dialysis, Hyogo College of Medicine Nishinomiya 663, Japan

[b] First Department of Medicine, Osaka University Medical School, Suita 653, Japan

INTRODUCTION

During antidiuresis, the mammalian kidney accumulates NaCl and urea in the papillary interstitium by accelerating the transport of sodium and the reabsorption of water. Medullary cells are exposed to interstitial hyperosmolality with high salts and urea, which they generate. These cells preferentially accumulate organic osmolytes over inorganic ions in response to extracellular hyperosmolality (1, 2). Osmotically active organic solutes protect the cellular macromolecules and processes from the perturbation of salts and urea.

While medullary organic osmolytes have been associated with urinary concentration, the relationship between organic osmolytes and a urinary concentrating defect has not been well clarified. In kidney in situ, medullary hypertonicity is constructed by medullary cells themselves. So, delayed or decreased accumulation of osmolytes in the renal medulla may be the cause of renal concentrating defects. In the present study, we chose two models of renal concentrating defect, i.e. potassium depletion and protein deprivation.

In potassium (K) depletion, a renal concentrating defect is a prominent and consistent finding. Several defects associated with urinary concentrating ability have been reported, including prostaglandin overproduction (3), blunted responsiveness to antidiuretic hormone, reduced papillary solute content, abnormal medullary oxidative metabolism, altered ADH release, and primary polydipsia. Decreased accumulation of organic osmolytes in renal medulla could explain some of these abnormalities associated with the concentrating defect.

The ability of kidneys to excrete concentrated urine is impaired when protein is omitted from the diet, and is improved when protein is included in the diet (4, 5, 6). It has been suggested that the action of protein is mediated primarily by the passive accumulation of urea in medullary interstitial fluid. Protein intake cannot be dissociated from urea handling by the kidney, as protein consumption increases excretion of nitrogenous waste and promotes water conservation (4). However, increased urine concentrating ability by protein feeding should be discriminated from the rise in urine osmolality by the addition of urea to urine. The exact mechanism by which protein deprivation impairs the ability to concentrate urine is not understood.

To clarify the relationship between organic osmolytes and urine concentration defects during potassium depletion or protein deprivation, we conducted the following experiments. First, we established that some organic osmolytes were decreased in the inner medulla of the kidney after 3 days of water deprivation. In potassium depleted-rats, we focused on sorbitol, a principal organic osmolytes in the inner medulla, and on aldose reductase. Sorbitol is synthesized from glucose, a reaction that is catalyzed by aldose reductase. In protein-deprived rats, medullary myo-inositol content and the level of sodium dependent myo-inositol transporter (SMIT) mRNA were mainly investigated. A protocol of 72-hr water deprivation was designed to study the accumulation of osmolytes at the maximal urinary concentration. From these experiments, we could not determine whether the decrease in organic osmolytes impaired urinary concentration or whether reduced medullary osmolality decreased the need for organic osmolytes. Therefore, we examined the effect of potassium depletion on organic osmolyte content after infusion of hypertonic sodium or potassium, or the effect of protein deprivation after infusion of hypertonic sodium and/or urea for 48 hours. The hypertonic sodium, potassium and/or urea infusion resulted in an equal amount of these solutes in the renal medulla and raised extracellular osmolality.

METHODS

Male Wistar rats (Nihon Charles River, Japan) were used in these experiments. Animals were pretreated with feeding low protein diet for protein-deprived model or low potassium diet for potassium-depleted model. The inferior vena cava was cannulated via a polyethylene catheter (PE 10 tube) inserted into the femoral. In each model, animals were received continuous infusion of the solution below.

Model / Protocol	Food	Water	Infusion	Period
Both				
NO HYPER	Deprived	Free access	Sham	72 hrs
DEHY	Deprived	Deprived	Sham	72 hrs
Potassium-depleted				
Na-INF	Deprived	Deprived	3.5 % NaCl	48 hrs
K-INF	Deprived	Deprived	3.2 % KCl	48 hrs
Protein-deprived				
Na-INF	Deprived	Deprived	3.5 % NaCl	48 hrs
UR-INF	Deprived	Deprived	12 % urea	48 hrs
Na + UR-INF	Deprived	Deprived	2.5 % NaCl +9% urea	48 hrs

Kidney tissue extracts were prepared using a method previously described (7). Betaine, GPC, sorbitol, myo-inositol and taurine were measured in tissue extracts by high-performance liquid chromatography (HPLC) (7). Aldose reductase activity was assayed spectrophotometrically in the homogenates from inner medullary tissues.

Total RNA was isolated from inner medullary tissues from one kidney and Northern analysis of aldose reductase and SMIT mRNA or slot blot analysis of SMIT mRNA were performed as described previously (8, 9).

RESULTS AND DISCUSSION

Potassium depleted rats

After water deprivation, the maximal urine concentration was tested. Urine osmolality and potassium concentration as well as medullary potassium and urea content were significantly lower in potassium-depleted rats than in control rats. Inner medullary contents of all organic osmolytes measured, i.e. myo-inositol, betaine, taurine, sorbitol, and GPC, were significantly lower in potassium-depleted rats than in control rat (Fig 1). We also demonstrated that a decrease in sorbitol was caused by a reduction in aldose reductase mRNA abundance in potassium-depleted rats that were subjected to water deprivation. The urinary concentrating defect was associated with a reduction in the medullary contents of these organic osmolytes.

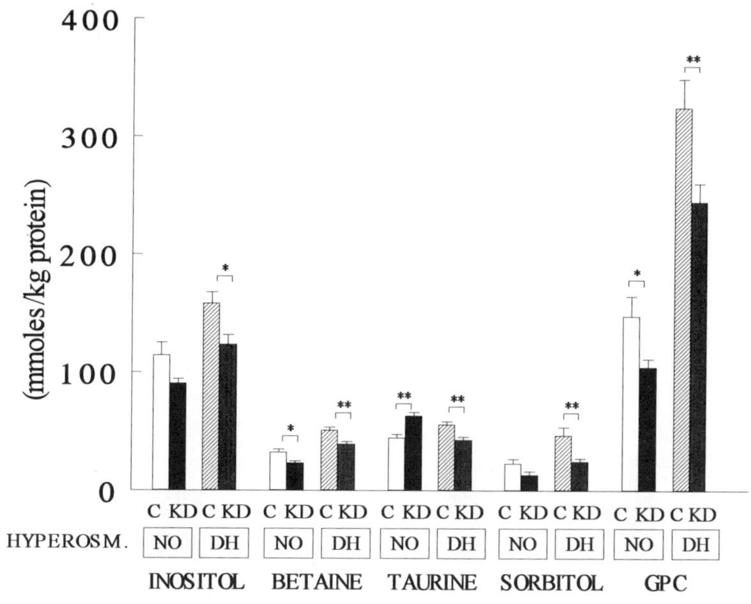

Fig 1 . Effect of potassium depletion on medullary contents of organic osmolytes during dehydration.
Values are means ± SE. Abbreviations are: C, control; KD, potassium depletion; NO, no hyperosmosis; DH, dehydration.
* P < 0.05, ** P < 0.01 significantly different vs. control

In Na infusion protocol, levels of sodium, potassium, and urea in the urine as well as urine osmolality were comparable to the findings in our previous studies (7). Urine potassium concentration were decreased in potassium-depleted rats, and the medullary

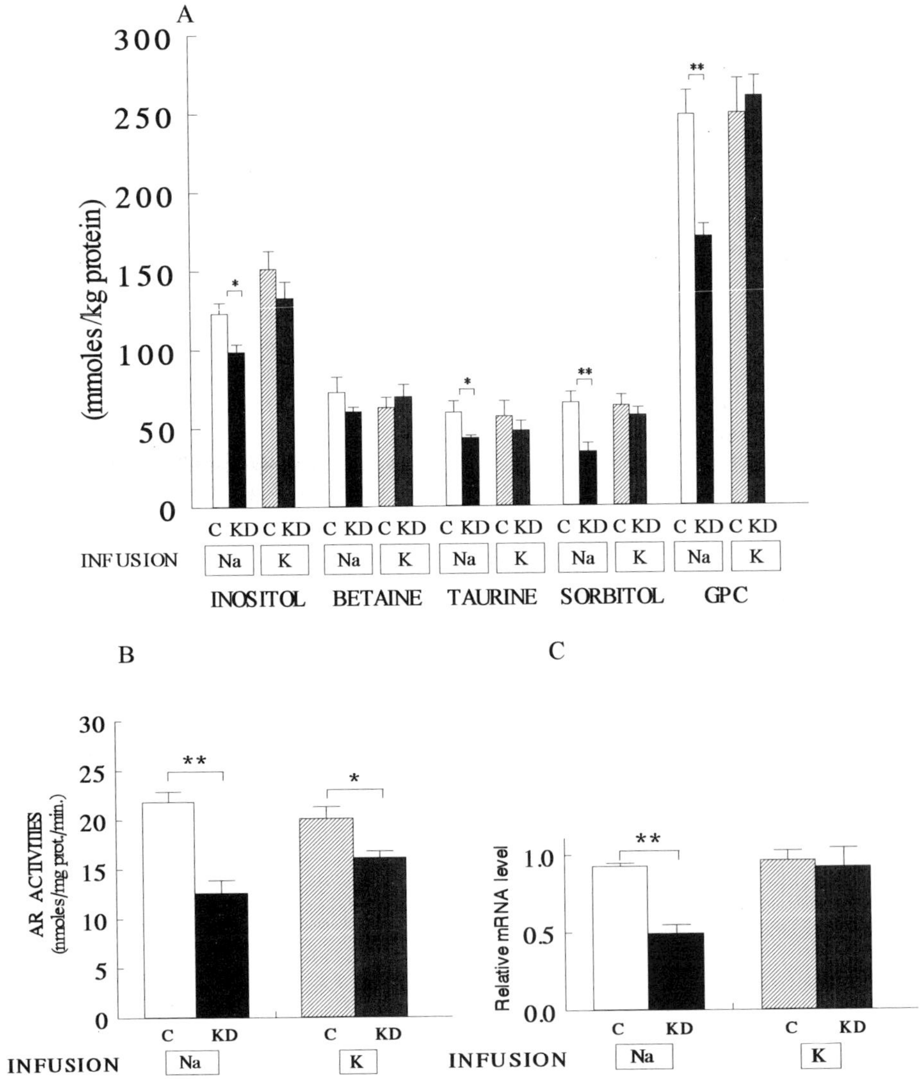

Fig 2. Effect of potassium depletion on medullary contents of organic osmolytes (A), aldose reductase (AR) activity (B), and the level of aldose reductase mRNA(C) during infusion of sodium or potassium. For the analysis of each blot, the relative change to the pooled RNA from dehydrated rats in densitometry scores were calculated.
Values are means ± SE. Abbreviations are: C, control; KD, potassium depletion.
* $P < 0.05$, ** $P < 0.01$ significantly different vs. control

contents of sodium and urea were significantly lower in the potassium-depleted vs. the control rats. Medullary contents of sorbitol, myo-inositol, taurine, and GPC were significantly lower in the potassium-depleted rats vs. the control (Fig 2A). Aldose reductase activity and mRNA abundance were all significantly lower in potassium-depleted rats even when they received an equal load of NaCl (Fig 2B, C).

Following the potassium infusion in potassium-depleted rats, the medullary content of sodium was decreased compared to control rats infused with potassium. No significant differences in the medullary contents of all organic osmolytes including sorbitol were observed (Fig 2A). The infusion of KCl for 48 hours restored sorbitol content and the aldose reductase mRNA abundance to the level in controls, but the activity of aldose reductase was significantly decreased in potassium-depleted rats, although it appeared to be higher than the protocol of NaCl infusion (Fig 2B, C). We don't know the exact mechanism of the discrepancy between enzyme activity and message level or sorbitol content. Not only the synthesis and degradation of sorbitol but also those of aldose reductase should be taken into consideration.

A relationship between potassium and the expression of osmoregulatory proteins has been demonstrated in mammalian cells and bacterial cells (10,11). In PAP-HT 25 cells (rabbit papillary cells) exposed to a medium made hyperosmotic by the addition of NaCl, there is an accumulation of sorbitol due to an elevation in the levels of aldose reductase mRNA and enzymatic activity. Extracellular Na and Cl need not be elevated because hyperosmolal raffinose in the medium is as effective as NaCl. The raffinose remains extracellular and causes the cells to shrink with a resultant elevation in the concentration of the potassium and sodium salts within the cells. Smardo and Burg et al. suggest that the intracellular concentrations of sodium plus potassium (or the intracellular ionic strength) may signal changes in the rate of aldose reductase transcription in PAP-HT25 cell (10). In the bacteria, Escherichia coli and Salmonella typhimurium, it was demonstrated that the ionic composition of the intracellular milieu, especially K^+ concentration can regulate gene expression of proU gene which is related to betaine transport for osmoprotection (11).

In the present study, we did not measure directly the intra- and extracellular concentrations of sodium and potassium, and medullary potassium content did not differ significantly between control and potassium-depleted rats in both infusion protocols. We could not determine directly the relation between intracellular electrolyte and organic osmolytes. However, it is reasonable to assume that sodium is mainly extracellular, potassium is the principal intracellular cation, and potassium infusion restore the intracellular potassium concentration. Therefore, we presumed that intracellular potassium or electrolytes linked to the level of aldose reductase mRNA, the upstream of sorbitol synthesis, in vivo.

Protein-deprived rats

After water deprivation, the maximal urine concentration was tested. Urine osmolality and urea concentration as well as medullary sodium and urea content were significantly lower in protein-deprived than in control rats. The increase in urine urea concentration, however, was as much in protein-deprived rats as in control. Inner medullary contents of myo-inositol as well as GPC and taurine were significantly lower in protein-deprived than in control rat (Fig 3). The urinary concentrating defect was associated with a reduction in the medullary contents of these organic osmolytes. These data suggest that the renal concentrating defect associated with protein deprivation may be due at least in part to an imbalance of osmolyte accumulation.

NaCl and/or urea infusion was used to load an equal amount of these solutes onto the kidney. The increased extracellular hyperosmolality caused by these solutes in the urine and medullary tissues might be expected to increase intracellular osmolyte in renal medullary cells, as in cultured cell models, cell osmolyte content is dependent on the osmolality of the medium (12).

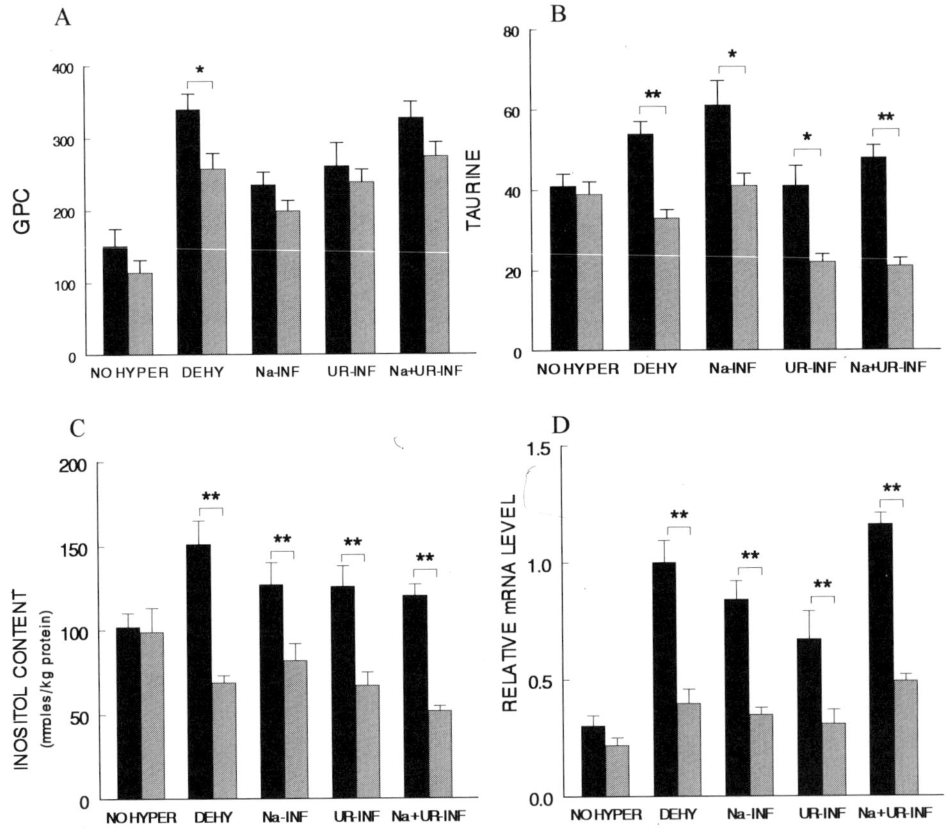

Fig 3. Effect of protein deprivation on the contents of myo-inositol and the level of SMIT mRNA in the renal inner medulla, during dehydration, or infusion of NaCl and/or urea. Each blot was analyzed by densitometry and the amount of RNA was normalized to that present in a standard obtained by pooling RNA from dehydrated rats.
Values are means ± SE. Solid bars, control group; hatched bars, protein-deprived group. Abbreviations are: NO HYPER, no hyperosmosis (untreated); DEHY, water deprivation; Na-INF, NaCl infusion; UREA-INF, urea infusion; Na+UREA-INF, NaCl plus urea infusion.
* P < 0.05, ** P < 0.01 significantly different vs. control.

NaCl infusion restored urinary Na concentration and medullary Na content to the level of control rats, but did not affect urine or medullary urea. The level of SMIT mRNA as well as inner medullary contents of myo-inositol and taurine were significantly lower in protein-deprived than in control rats.

Urea infusion did not increase the level of SMIT mRNA to that found in control rats, in which a decrease in myo-inositol content also was detected, although medullary urea content increased to the level of control rats.

NaCl plus urea infusion increased urinary Na and urea concentrations and medullary sodium contents to levels observed in control rats. Protein deprivation, however, impaired the rise in medullary urea, which was associated with a reduction in medullary contents of myo-inositol and taurine as well as SMIT mRNA. Thus, the combined administration of NaCl and urea did not restore SMIT mRNA abundance or myo-inositol or taurine content in protein-deprived rats.

In protein deprived animals, the rise in Na and/or urea in renal medulla and urine failed to increase myo-inositol and taurine contents to the levels present in control rats (Fig 3). The failure to increase the level of SMIT mRNA by any hyperosmolality protocol accounted for the decreased medullary content of myo-inositol. A factor associated with protein feeding other than NaCl and urea might be responsible for the renal medullary accumulation of myo-inositol and taurine and the impairment of urine concentrating ability.

SUMMARY

The present study suggested that a decrease in organic osmolytes might precede the impairment of urinary concentration in both potassium-depleted and protein-deprived models. The decrease in intracellular osmolality associated with reduced organic osmolytes in the renal medulla might be correlative with renal concentrating defect.

From these observations, we hypothesized the mechanism of urinary concentrating defect in collecting tubules. Fig 4A shows the normal dehydrated condition and Fig 4B is

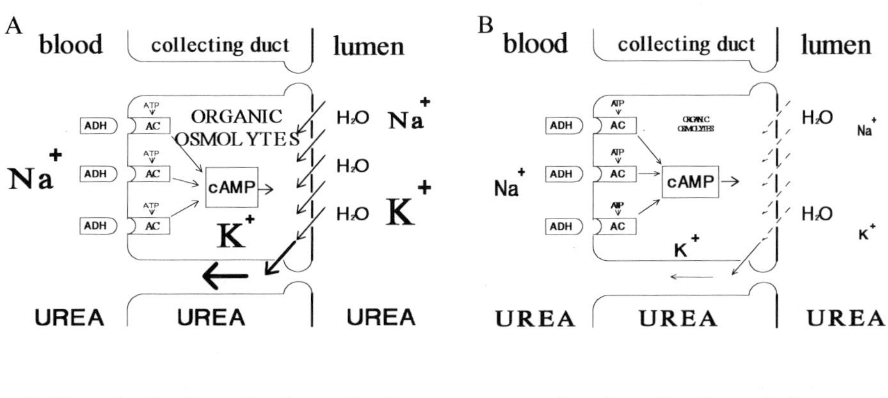

Fig 4. Hypothesized mechanism of urinary concentration in collecting tubules: control (A) and potassium-depletion (B). Abbreviations are; ADH, antidiuretic hormone; AC, adenylcyclase

the dehydrated condition with potassium depletion. In the collecting tubules, ADH increases water permeability by the insertion of water channels in the apical cell membrane. Water would move from the apical side to the hypertonic interstitium. If there are not sufficient osmotic substances in these cells, then water would not pass the cell membrane. As intracellular osmolality could be composed of organic osmolytes as well as potassium, reduced organic osmolytes accumulation in the renal medulla might impair the renal concentrating ability.

Therefore, we presumed that the initial but small increase in medullary tonicity associated with Na reabsorption in the loop of Henle accelerated regulatory volume increase and the uptake of potassium into the cells of collecting tubules, when the body was dehydrated. In these cells, the rise in intracellular ionic strength may stimulate organic osmolytes accumulation to maintain the osmotic balance between the inside and outside of the cells. Failure to accumulate organic osmolytes in collecting tubules might attenuate water reabsorption and derange urine concentration.

REFERENCES

1. Yancey PH, Clark ME, Hand SC, Bowlus RD, Somero GN. Living with water stress: Evolution of osmolyte system. Science 1982; 217:1214-1222.
2. Garcia-Perez A, Burg MB. Renal medullary organic osmolytes. Physiol Rev 1991; 71:1081-1115.
3. Raymond KH, Lifschitz MD, and McKinney TD. Prostaglandins and the urinary concentrating defect in potassium-depleted rabbits. Am. J. Physiol., 1987; 253:F1113-1119.
4. Bankir L. Urea and the kidney. In: Brenner, BM eds. The kidney. Saunders, Philadelphia 1996; 571-606.
5. Bray GA, and Preston AS. Effect of urea on urine concentration in the rat. J. Clin. Invest. 1961; 40: 1952-1960.
6. Epstein FH, Kleeman CR, Pursel S, and Hendrikx A. The effect of feeding protein and urea on the renal concentrating process. J. Clin. Invest. 1957; 36: 635-641.
7. Nakanishi T, Uyama O, Sugita M. Osmotically regulated taurine content in rat renal inner medulla. Am J Physiol 1991; 261:F957-F962.
8. Nakanishi T, Nishihara F, Yamauchi A, Yamamoto S, Sugita M, Takamitsu Y. NaCl and/or urea infusion fails to increase renal inner medullary myo-inositol in protein-deprived rats. Am J Physiol 1996; 271:F1255-F1263.
9. Nakanishi T, Yamauchi A, Yamamoto S, Sugita M, Takamitsu Y. Potassium depletion modulates aldose reductase mRNA in rat renal inner medulla. Kidney Int 1996; 50:828-834.
10. Smardo FLJr, Burg MB, Garcia-Perez A. Kidney aldose reductase gene transcription is osmotically regulated. Am J Physiol 1992; 262:C776-C782.
11. Sutherland L, Cairney J, Elmore MJ, Booth IR, Higgins CF. Osmotic regulation of transcription: Induction of proU betaine transport gene is dependent on accumulation of intracellular potassium. 1986; J Bacteriol 168:805-814.
12. Nakanishi T, Balaban RS, Burg MB. Survey of osmolytes in renal cell lines. Am. J. Physiol. 1988; 255: C181-C191.

Cell Volume Regulation: The Molecular Mechanism and Volume Sensing Machinery
Y. Okada, editor.

HYPOTONIC AND HIGH K$^+$ MEDIA SWELLING-INDUCED RELEASE OF EXCITATORY AMINO-ACIDS FROM BRAIN ASTROCYTES

H. K. Kimelberg[a,b,c] and E. M. Rutledge[a]

[a]**Department of Pharmacology/Neuroscience Albany Medical College, Albany, New York 12208**

[b]**Neuropharmacology and Neuroscience Research Group Albany Medical College, Albany, New York 12208**

[c]**Division of Neurosurgery, Albany Medical College, Albany, New York 12208**

SUMMARY

During stroke or head trauma extracellular potassium concentrations increase which can cause both astrocytes and neurons to swell. The excitatory amino acids (EAAs) released by non-exocytotic mechanisms because of cellular swelling may contribute to excitotoxicity *in vivo*. Several putative swelling-activated channels have been identified through which such cellular osmolytes can be released. In the present study we sought to identify the swelling activated channel(s) responsible for [^{3}H]-D-aspartate release from primary cultured astrocytes exposed to raised potassium or hypotonic medium. This high KCl-induced [^{3}H]-D-aspartate release was inhibited by extracellular ATP and tamoxifen. Tamoxifen also inhibited hypotonic media-induced release, but extracellular ATP had no effect. KCl-induced cell swelling was not inhibited by extracellular ATP or tamoxifen, which suggests that these anion channel inhibitors directly blocked the KCl-induced swelling-activated channel. Of the swelling-activated channels so far identified our data suggest that the volume sensitive osmolyte anion channel or VSOAC (also known as volume sensitive outwardly rectifying channel or VSOR) is responsible for [^{3}H]-D-aspartate release during raised extracellular K$^+$, and probably responsible for [^{3}H]-D-aspartate release during hypotonic-swelling induced release from primary cultured astrocytes.

BACKGROUND

Most animal cells release free cellular inorganic and organic cations and anions when swollen by exposure to hypotonic media, which enables them to regulate their cell volumes back to normal, a process called regulatory volume decrease or RVD. In mammalian cells the main inorganic osmolytes lost released during hypotonic-induced swelling are K$^+$ and Cl$^-$ which probably contribute the most to RVD due to their high intracellular concentrations [1,2]. The main organic osmolytes lost are the amino acids and their derivatives. These include taurine, a major osmoregulator which is preferentially lost during cell swelling [3], and to a lesser extent glutamate, aspartate, GABA, and alanine [4,5].

In various brain pathologies such as stroke and traumatic brain injury (TBI) there is marked brain and cellular swelling which is categorized into vasogenic and cellular (cytotoxic) edema, respectively. Cellular edema is fundamentally due to a shift of osmolytes and brain water from the extracellular space into cells, while vasogenic edema refers to an increase in net brain volume [6].

Large changes in extracellular K^+, reaching concentrations of up to 80 mM, have been measured with extracellular potassium electrodes *in vivo* in cerebral ischemia and trauma [7,8]. Such increases in extracellular K^+ might be a key contributor to cellular edema during pathological conditions in the CNS, since raised extracellular K^+ causes cell swelling due to Donnan diffusional forces [9,10].

In many cultured cells, including primary astrocyte cultures, release of organic osmolytes is believed to be mediated by diffusion through swelling-activated anion channels, rather than on co-transporters or exchangers [5,11,12,13,14]. Four types of swelling-activated anion channels have so far been identified [13,14]. 1) ClC-2 and ClC-3, which both belong to a family of channels designated as ClC. However, they have very different I-V curves, showing inwardly and outwardly rectifying currents, respectively. Their pharmacology is also different with ClC-2 being inhibited by diphenylamine carboxylate whereas ClC-3 is inhibited by a variety of anion channel blockers such as DIDS, millimolar concentrations of extracellular ATP and 5-nitro-2-(3-phenylpropylamino) benzoic acid (NPPB) [13,14]. 2) A swelling-activated channel termed the 'maxi' anion channel which has been identified in cultured cortical astrocytes, neuroblastoma cells and in the apical membrane of renal collecting duct cells. It shows multiple activation states with a large unitary conductance of 200-400 pS upon hypotonic-induced swelling. It is inhibited by DIDS, L-644,711, NPPB, pertussis toxin and inhibitors of protein kinase C [13,14]. 3) A volume sensitive organic anion channel or VSOAC [17], also termed a volume expansion sensitive outwardly rectifying or VSOR channel, which is inhibited by extracellular 1-10 mM ATP, 100 μM NPPB and 10 μM tamoxifen [13,14]. 4) A channel termed I_{Cln}, which has been cloned [15], has similar characteristics to the VSOAC.

MAIN QUESTIONS

Our overall objective is to elucidate the mechanisms and significance of increased extracellular excitatory amino-acids (EAAs) released by the swelling of astrocytes in the brain during and after cerebral ischemia. Increased extracellular $[K^+]$ replacing Na^+ with $[Cl^-]_o$ kept constant, which we refer to as elevated KCl, has been shown to stimulate [^{3}H]-taurine and [^{3}H]-D-aspartate release from primary cultures of mouse and rat primary astrocyte cultures, respectively [3,4,11,16]. We have used the pharmacology of swelling-activated anion channels, noted above, to identify the channel(s) responsible for the KCl- and hypotonic media-induced release of [^{3}H]-D-aspartate from primary cortical astrocyte cultures.

SUMMARY OF OUR WORK

Culture methods

Primary astrocyte cultures were prepared from the cerebral cortex as previously described [17]. Immunocytochemistry showed that > 95% of the cells stained positively for the astrocytic marker, glial fibrillary acidic protein (GFAP). The dissociated cells were seeded and grown on poly-D-lysine coated 18 x 18 mm coverslips (Bellco Biotechnology, Vineland, NJ). The cultures were used when the cells formed a confluent monolayer after approximately 3-4 weeks.

Efflux Measurements

Astrocytes on the coverslips were incubated overnight in 2.5 ml of MEM containing 10% horse serum together with 4 μCi/ml of [^{3}H] D-aspartate, which was used as a non-metabolizable marker for the behavior of intracellular glutamate and aspartate pools [18]. To measure efflux the coverslips were inserted into a laboratory-designed Lucite chamber and perfused at 37° with HEPES-buffered solutions consisting of 140 mM NaCl, 3.3 mM KCl, 0.4 mM $MgSO_4$, 1.3 mM $CaCl_2$, 1.2 mM KH_2PO_4, 10 mM D (+) glucose, 25 mM HEPES, pH 7.4. 100 mM KCl buffers were made by replacing Na^+ with equimolar K^+ . The volume above the cells in the shallow perfusion chamber was approximately 50 μl and the perfusion rate was ~1 ml/min. This small volume gives us high precision in our efflux studies and allows the measurement of electrical resistance changes reflecting average cell volume changes, as described below. Release for each time point was calculated by dividing the radioactivity released in a one minute sample by the total radioactivity which would have been present at the beginning of that time point as calculated from the sum of the subsequent one minute fractions plus the amount present in the cell digest at the end of the experiment. We term this "percent fractional release".

Cell Volume measurements

KCl-induced swelling was measured using an electrical resistance method, based on principles previously described [19]. Briefly, a lock-in amplifier supplies an alternating 5 volt signal at 500 Hz through a 1 MΩ resistor in series with the chamber. Because the resistance of the perfusion chamber and the rest of the current path is small relative to this series resistance this gives an essentially constant current of 5 μA. The same Lucite chamber was used as in the perfusion experiments. Cell shrinkage or swelling produces an increase or decrease in the volume of medium above the cells and therefore a decrease or increase in the electrical resistance of the chamber, respectively. Voltage across the chamber is thereby altered and detected by the amplifier system. Thus, small changes in resistance that result from changes in the astrocyte monolayer volume are measured.

Increased medium KCl shows two components of [^{3}H]-D-aspartate release: later component is inhibited by ATP.

The first response in Figure 1A shows a typical [^{3}H]-D-aspartate release in response to a 20 minute exposure to a 100 mM KCl HEPES buffer in which Na^+ was replaced by K^+. There was usually, but not always, a small initial transient peak (see also Figures 2A,B) followed by a slower, progressively increasing, second phase of release which rapidly returned to baseline when the perfusate was returned to the normal 3.3 mM K^+. We have previously shown that the initial transient peak represents reversal of the high affinity glutamate uptake transporter and can be greatly enhanced by pretreatment with ouabain to increase $[Na^+]_I$, a condition which mimics ischemia. The larger, progressively increasing, second phase of release is due to cell swelling [20]. The second exposure to KCl in Figure 1A was with 5 mM ATP, an inhibitor of the VSOAC channel, also present throughout the 100 mM KCl exposure. ATP caused a rapid initial increase in [^{3}H]-D-aspartate release but this release plateaued and remained constant until the media was switched back to normal $[K^+]_o$. The stimulating response was due to an effect of ATP itself because the same concentration of ATP in isotonic media also caused stimulation of [^{3}H]-D-aspartate release (data not shown). The subsequent plateau of the release suggests that the progressively increasing second phase of [^{3}H]-D-aspartate release was inhibited.

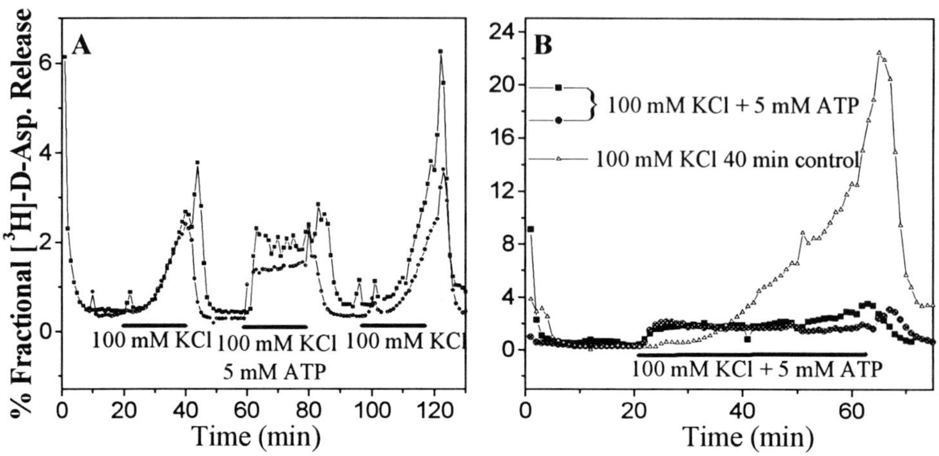

Figure 1 The effects of 5 mM extracellular ATP on 100 mM KCl-induced [^{3}H]-D-aspartate release. A: Three consecutive 20 min 100 mM KCl exposures with the middle KCl exposure containing 5 mM ATP. There is stimulation of release caused by ATP which then plateaued (two separate experiments shown as ● and ■). B: A longer 40 min 100 mM KCl exposure in the absence of 5 mM ATP (Δ) and two separate experiments in the presence of 5 mM ATP (● and ■).

Since the effects were complicated by the rapid stimulation of release due to ATP we reasoned that the effect of extracellular ATP on the second component might be more apparent over a longer period of exposure to 100 mM KCl. We therefore doubled the exposure time to increased KCl. Figure 1B shows that 5 mM extracellular ATP again stimulated release (note larger y axis scale in Figure 1B compared to Figure 1A), but after a 16 min exposure the KCl control surpassed the release where 5 mM ATP was also present and continued to increase for the remaining 24 minutes. The inhibition by 5 mM ATP of the second, swelling-induced phase of release is now much clearer.

The effects of tamoxifen and NPPB on raised K^+-induced [^{3}H]-D-aspartate release.

Tamoxifen is another specific inhibitor of the VSOAC compared to other anion channels [13,14]. It has been shown to inhibit both taurine and ^{125}I efflux in HeLa cells with an IC_{50} of approximately 1-2 μM [21]. In Figure 2A we show that tamoxifen at 15 μM inhibited KCl-induced release of [^{3}H]-D-aspartate release by 63 and 92 % in the two experiments shown. The tamoxifen effect is partially irreversible so we do not show a third response.

NPPB (5-nitro-2-(3-phenylpropylamino)-benzoic acid) is an anion channel inhibitor that inhibits a number of anion channels including the VSOAC, the maxi, and ClC-3 channels. In Figure 2B we show that 100 μM NPPB completely inhibited KCl-induced release. This effect was reversible as shown in the third 100 mM KCl exposure in the absence of NPPB.

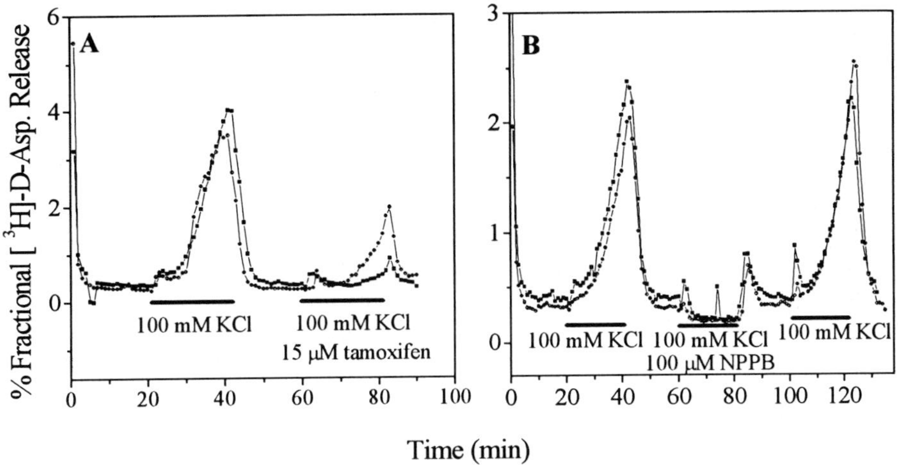

Figure 2. The effect of tamoxifen and NBBP on KCl-induced [^{3}H]-D-aspartate release. A: Two consecutive 20 min 100 mM KCl exposures in 2 experiments (O and ■) with 15 μM tamoxifen present during the second exposure. Tamoxifen inhibited [^{3}H]-D-aspartate release by 65 and 92 % compared to first control exposure in the two experiments shown. B: 100 μM NPPB reversibly and completely inhibited 100 mM KCl-induced release in the second exposure to 100 mM KCl (two separate experiments are shown, (O and ■). The same percentage of DMSO or ethanol used as solvents were present in all media with or without inhibitors. There was no detectable effect of DMSO or ethanol alone.

The effects of NPPB and tamoxifen on hypotonic media-induced [^{3}H]-D-aspartate release

The time course of hypotonic media-induced [^{3}H]-D-aspartate release and swelling is very different from K^+-induced swelling and release. Hypotonic-media-induced [^{3}H]-D-aspartate release shows an immediate and rapid rate of [^{3}H]-D-aspartate release which peaks after 3-5 minutes and then declines. This decline in [^{3}H]-D-aspartate release, or other amino-acids such as taurine, in hypotonic medium mirrors cell volume regulation (RVD) and presumably reflects the progressive closure of channels as the cells restore normal volume [12,21]. Figure 3A shows that 100 μM NPPB inhibited hypotonic-induced release by ~ 90 % and Figure 3B shows that 15 μM tamoxifen inhibited by ~ 75 %. However, 5-10 mM extracellular ATP had no effect (data no shown).

The effect of inhibitors on KCl-induced swelling.

We have used elevated KCl-induced media to induce swelling because this is a likely effector of astrocytic swelling in pathological states *in vivo* (see introduction). However, such swelling involves uptake of KCl followed by osmotically obligated water and therefore anion channel blockers could inhibit such swelling by blocking Cl^- influx. We therefore examined the effects of the anion transport inhibitors we used on KCl-induced swelling of cultured astrocytes as measured by the electrical resistance method (see Methods). As shown in Figure 4 the time course for KCl-induced swelling was similar to the second phase of [^{3}H]-D-aspartate release in that swelling also progressively increased for as long as the raised KCl medium was present. As shown in Figure 4 extracellular ATP at 10 mM had no effect on KCl-induced swelling. Tamoxifen (15μM) had no effect over the initial 30 min of 100 mM KCl but after 40 min the rate of swelling appeared to slow. NPPB treated cells shrunk initially followed by a small increase in cell volume. The baseline for isotonic medium was a horizontal line and essentially equivalent to the line shown for room temperature (20-22°C).

Volume Regulatory Mechanisms and high K^+ -induced swelling and osmolyte release.

The fact that extracellular ATP and tamoxifen inhibit KCl induced [^{3}H]-D-aspartate efflux (Figures 1 and 2) but not KCl-induced swelling (Figure 4), allows us to conclude that such efflux is likely to be through the VSOAC / VSOR channel. The initial stimulation of [^{3}H]-D-aspartate release by ATP appears to be independent of KCl because stimulation was also seen in isotonic medium (data not shown). ATP might be activating a non-specific channel via activation of P_{2y} receptors on cultured astrocytes which allows taurine and EAAs to pass through. Stimulation of reversal of the glutamate transporter is unlikely because the ionic gradients in isotonic media do not favor efflux [20]. Taurine efflux only appears to be mediated by a channel-like mechanism [12].

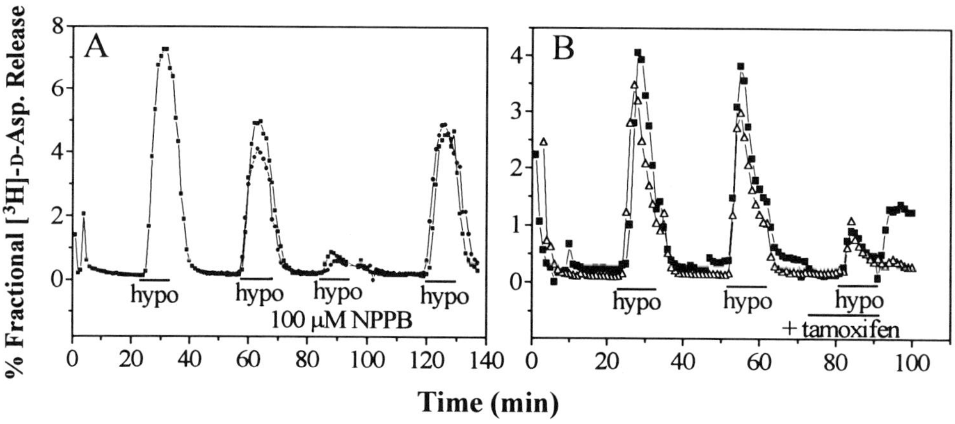

Figure 3. The effect of NPPB and tamoxifen on hypotonic-induced [^{3}H]-D-aspartate release. **A:**(● and ■) 100 μM NPPB completely inhibited hypotonic-induced release in the third exposure. Hypotonic media was made by removing 50 mM NaCl (-100 mOsmol). A 0.1 % DMSO/ethanol (50/50) was present throughout the NPPB experiment. **B:** Fifteen min after the second hypotonic exposure the cells were pretreated with 15 μM tamoxifen for 10 min prior to the third hypotonic exposure when 15 μM tamoxifen was also present (■ and Δ). The same percentage of ethanol (0.04 % v/v) was present throughout the experiment as with tamoxifen containing media. A fourth hypotonic exposure was not done because previous results showed tamoxifen to be irreversible.

One feature of KCl-induced swelling is the absence of apparent RVD. This is probably the result of K^+ and Cl^- contributing most to RVD in cultured astrocytes during hypotonic exposure, and in raised potassium media net K^+ efflux is reduced or eliminated and amino acids are not in high enough concentration to counteract the swelling [3]. However, if one could specifically block the KCl efflux pathways one might see that the swelling is actually a steady state balance between influx of KCl and efflux of KCl and other osmolytes.

Pharmacology of hypotonic media-induced release of [^{3}H]-D-aspartate.

The rapidity and magnitude of hypotonic-media swelling can probably activate a variety of swelling-activated anion channels. An example of this is non-pigmented epithelial retinal cells which show three separate anion channel conductances when swollen by hypotonic media; low (6.7 pS), intermediate (18 pS) similar to the VSOAC/VSOR, and large conductance (100 pS) anion channels. NPPB at a concentration of 100 μM blocked all three currents [22]. Since the hypotonic-media-induced [^{3}H]-D-aspartate release is insensitive to extracellular ATP this would exclude the VSOAC, whereas NPPB completely blocked release. However, tamoxifen is considered to be a specific inhibitor of VSOAC/VSOR channels and therefore supports that [^{3}H]-D-aspartate release is via the VSOR channel. In addition when cells are depleted of intracellular ATP, elevated K^+ and hypotonic-media induced [^{3}H]-D-aspartate release are both inhibited by 100 % (unpublished observation), which is another characteristic of the VSOAC/VSOR channel.

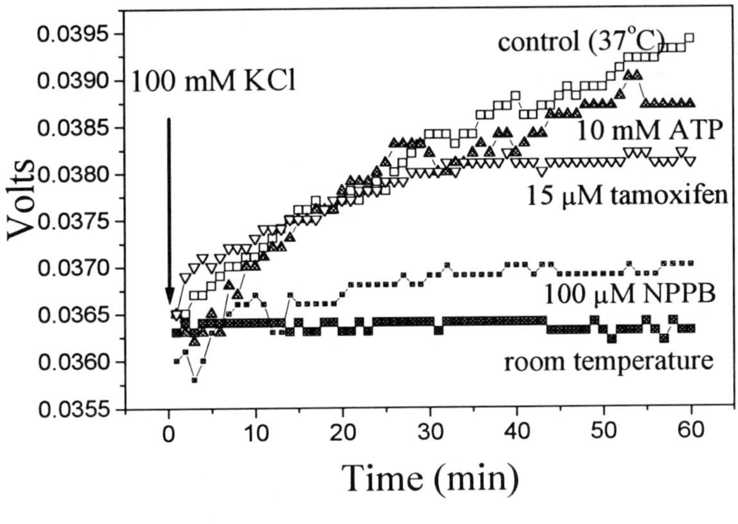

Figure 4. The effect of anion channel blockers and a reduction of temperature on KCl-induced swelling as measured by electrical impedance. 100 mM KCl was introduced at time 0 and continued for the entire 60 min shown. A voltage increase indicates an increased resistance reflecting a proportional decrease in the medium volume above the cells caused by cell swelling (Control, n = 5; ATP, n= 3; tamoxifen, n = 3; NPPB, n = 3 and room temperature, n = 3).

Hypotonic-media-induced [^{3}H]-D-aspartate release may be less sensitive to ATP because membrane depolarization during this type of swelling is less than the depolarization seen when cells are exposed to 100 mM K^+. The increased depolarization of the membrane potential in 100 mM K^+ can better drive ATP into the channel [23]. Therefore, these results suggest that the VSOAC/VSOR channel is involved for [^{3}H]-D-aspartate release regardless of the method used to cause cell swelling.

Other effects of the inhibitors also complicate interpretation. NPPB has previously been shown to be a metabolic uncoupler and can reduce ATP levels to < 50% of control values within 2 minutes [24]. The VSOAC/VSOR activation depends on intracellular ATP and is inhibited when intracellular ATP levels drop below 50 % of normal [25]. It was shown that NPPB did not directly block the volume-sensitive anion current suggesting that inhibition by NPPB of this current could be due to intracellular ATP depletion [24].

PERSPECTIVES

Astrocytes contribute substantially to the cellular swelling that occurs as a result of traumatic brain injury and ischemia, and may contribute to an unfavorable outcome due to the release of EAAs [6]. In an animal brain injury model inhibition of astrocytic swelling has been shown to be associated with improved neurological outcome and mortality [26]. The focus of our studies now is to identify in primary astrocyte cultures the swelling-activated channel(s) possibly responsible for such excitatory amino acid release, using raised KCl to swell the cells. Since elevated extracellular K^+ is seen in ischemia [7,8] we also wish to test if release of EAAs by such a mechanism occurs *in vivo* in cerebral ischemia and if inhibition by specific channel blockers confers neuroprotection. So far we have seen that an anion channel inhibitor partially blocks ischemia-induced release of glutamate and aspartate during ischemia in rat striatum as measured by microdialysis [27]. Another study reported that several anion transport inhibitors reduced ischemia-induced glutamate and aspartate release, measured using a cortical cup perfusion system over the exposed cortex [28]. By obtaining a better understanding of which type of swelling activated channel(s) contribute to EAA release during brain cellular edema more specific drugs can be designed to block these channels which may be therapeutically useful.

ACKNOWLEDGEMENTS

Supported by NIH grant NS35205 awarded to HKK. We thank Dr. Michael Aschner, Bowman Gray School of Medicine, for supplying the data shown in Figure 3.

BIBLIOGRAPHY

1. Gilles, R., E. K. Hoffman, and L. Bolis. *Advances in Comparative and Environmental Physiology, vol 9 - Volume and Osmolarity Control in Animal cells*. Berlin: Springer-Verlag, 1991.

2. Hoffman, E. K. and L. O. Simonsen. Membrane mechanisms in volume and pH regulation in vertebrate cells. *Physiol. Rev.* 69: 315-382, 1989.

3. Pasantes-Morales, H., S. Alavez, R. Sánchez-Olea, and J. Morán. Contribution of organic and

inorganic osmolytes to volume regulation in rat brain cells in culture. *Neurochemical Res* 18: 445-452, 1993.
4. Kimelberg, H. K., S. K. Goderie, S. Higman, S. Pang, and R. A. Waniewski. Swelling-induced release of glutamate, aspartate, and taurine from astrocyte cultures. *J. Neurosci* 10: 1583-1591, 1990.

5. Roy, G. Amino acid current through anion channels in cultured human glial cells. *J. Membrane Biol.* 147: 35-44, 1995.

6. Kimelberg, H. K. Current concepts of brain edema: Review of laboratory investigations. *J. Neurosurg.* 83: 1051-1059, 1995.

7. Hansen, A. Effect of anoxia on ion distribution in the brain. *Physiol. Rev.* 65: 101-138, 1985.

8. Siesjö, B. K. Cerebral circulation and metabolism. *J. Neurosurg* 60: 883-908, 1984.

9. Hodgkin, A. L. and P. Horowicz. The influence of potassium and chloride ions on the membrane potential of single muscle fibres. *J. Physiol.* 148: 127-160, 1959.

10. Walz, W., A. Klimaszewski, and A. I. Paterson. Glial swelling in ischemia: a hypothesis. *Dev. Neurosci.* 15: 216-225, 1993.

11. Sánchez-Olea, R., C. Peña, J. Morán, and H. Pasantes-Morales. Inhibition of volume regulation and efflux of osmoregulatory amino acids by blockers of Cl^- transport in cultured astrocytes. *Neurosci Lett* 156: 141-144, 1993.

12. Schousboe, A., R. Sánchez Olea, J. Morán, and H. Pasantes-Morales. Hyposmolarity-induced taurine release in cerebellar granule cell is associated with diffusion and not with high-affinity transport. *J Neurosci Res* 30: 661-665, 1991.

13. Okada, Y. Volume expansion-sensing outward-rectifier Cl^- channel: fresh start to the molecular identify and volume sensor. *Am J Physiol* 273: C755-C789, 1997.

14. Strange, K., F. Emma, and P. S. Jackson. Cellular and molecular physiology of volume-sensitive anion channel. *Am. J. Physiol.* 270: C711-C730, 1996.

15. Paulmichl, M., Y. Li, K. Wickman, M. Ackerman, E. Peralta, and D. Clapham. New mammalian chloride channel identified by expression cloning. *Nature* 356: 238-241, 1992.

16. Martin, D. L., V. Madelian, B. Seligmann, and W. Shain. The role of osmotic pressure and membrane potential in K+-stimulated taurine release form cultured astrocytes and LRM55 cells. *J Neurosci* 10: 571-577, 1990.

17. Frangakis, M. V. and H. K. Kimelberg. Dissociation of Neonatal Rat Brain by Dispase for Preparation of Primary Astrocyte Cultures. *Neurochem Res* 9: 1689-1698, 1984.

18. Erecinska, M. and I. A. Silver. Metabolism and role of glutamate in mammalian brain. *Prog. Neurobiol.* 35: 245-296, 1990.
19. O'Connor, E. R., H. K. Kimelberg, C. R. Keese, and I. Giaever. Electrical Resistance Method for Measuring Volume Changes in Monolayer Cultures Applied to Primary Astrocyte Cultures. *Am. J. Physiol.* 264: C471-C478, 1993.

20. Rutledge, E. M. and H. K. Kimelberg. Release of [^{3}H]-D-aspartate from primary astrocyte cultures in response to raised external potassium. *J Neurosci* 16: 7803-7811, 1996.

21. Kirk, J. and K. Kirk. Inhibition of volume-activated I- and taurine efflux from HeLa cells by P-glycoprotein blockers correlates with calmodulin inhibition. *J Bio Chem* 269: 29389-29394, 1994.

22. Zhang, J. J. and T. J. Jacob. Three different Cl- channels in the bovine ciliary epithelium activated by hypotonic stress. *J. Physiol* 499: 379-389, 1997.

23. Jackson, P. S., K. Strange. Characterization of voltage-dependent properties of a voltage sensitive anion conductance. *J. Gen. Physiol.* 105 661-667, 1995.

24. Ballatori, N., A. T. Truong, P. S. Jackson, K. Strange, and J. L. Boyer. ATP depletion and inactivation of an ATP-sensitive taurine channel by classic ion channel blockers. *Mol Pharm* 48: 472-476, 1995.

25. Jackson, P. S., R. Morrison, and K. Strange. The volume sensitive organic osmolyte-anion channel VSOAC is regulated by nonhydrolytic ATP binding. *Am.. J. Physiol.* 267: C1203-C1209, 1994.

26. Barron, K. D., M. P. Dentinger, H. K. Kimelberg, L. R. Nelson, R. S. Bourke, S. Keegan, R. Mankes, and E. J. Cragoe. Ultrastructural feature of a brain injury model in cat. I. Vascular and neuroglial changes and the prevention of astroglial swelling by a fluorenyl(aryloxy) aldanoic acid derivative (L-644, 711). *Acta. Neuropath.* 75: 295-307, 1988.

27. Seki, Y., P. J. Feustel, H. K. Kimelberg, R. W. Keller, B. I. Tranmer, C. Charniga, A. Chandra, and A. J. Popp. Inhibition of an astrocyte glutamate transporter reduces ishemia-induced glutamate release in the striatum. *Soc. for Neurosci Abst.* 901.12, 1997.

28. Phillis, J. W., M. S. Smith-Barbour, L. M. Perkins, and M. H. O'Regan. Characteristics of Glutamate, Aspartate, and GABA Release from Ischemic Rat Cerebral Cortex. *Brain Res. Bull.* 34: 457-466, 1994.

Cell Volume Regulation: The Molecular Mechanism and Volume Sensing Machinery
Y. Okada, editor.

Structure and regulation of the aquaporin water channels

Kiyohide Fushimi[ab], Sei Sasaki[a], and Fumiaki Marumo[a]

[a]Second Department of Internal Medicine and [b]Department of Medical Informatics, Tokyo Medical and Dental University School of Medicine, 1-5-45 Yushima, Bunkyo-ku, Tokyo 113-8519, Japan

SUMMARY

Aquaporins found in many organisms are ancient membrane intrinsic proteins that provide a pathway for water molecules. They in common have six transmembrane segments and characteristic tandem repeat structure with an asparagine-proline-alanine (NPA) motif. Functions of aquaporins seem to be regulated by translocation and modification of channels. Some mutations in aquaporin genes cause diseases including nephrogenic diabetes insipidus and cataract, implicating physiological importance of the aquaporin gene family.

INTRODUCTION

The aquaporins (AQP) are a family of membrane channel proteins that confer high water permeability to cell membranes of animals, plants, and bacteria. (1). In mammals, nine aquaporins have been identified, so far, and the number of the aquaporin members seems to be still increasing. AQP1 is mainly in red blood cell membrane and kidney proximal tubule (2). AQP2 is the vasopressin water channel in luminal membrane of kidney collecting duct where vasopressin regulates osmotic water permeability (3). AQP3 is expressed in the kidney and colon (4), AQP4 in the brain, lung, and kidney (5), AQP5 in secretory glands (6), AQP6, a close homologue of AQP2, in the kidney, AQP7 (7) and AQP8 (8) mainly in testis. MIP has been shown to possess minimal water channel function (9), being designated as AQP0. Of nine aquaporins, AQP4 and AQP7 are resistant to mercury reagents; AQP3 and AQP7 are permeable to glycerol in addition to water. In plants, numerous aquaporins have been identified in the tonoplast membrane (AQP-TIPs) and in the plasma membrane (AQP-PIPs) (10, 11).

MOLECULAR STRUCTURE OF AQUAPORINS

Investigations have been conducted to identify the molecular structure of the aquaporins that accounts for selective permeability to water molecules. The localization and three-dimensional structure of the aqueous pathway of aquaporins have become elucidated by extensive examination of AQP1 structure using methods of molecular engineering and X-ray crystallography. Studies by Agre et al. has shown that AQP1 exists in the plasma membrane in a homo-tetramer form; that each monomer subunit has one aqueous pore; and that N-glycosylation moiety attaches to one of four subunits but is not necessary for water channel function (12). Hydropathy analysis predicted six-membrane-spanning topology, assuming α-helical conformation of transmembrane segments. Agre et al. proposed the hourglass model for the structure of the water pore (13), based on the findings from a series of site-directed mutagenesis studies. They found that mutations near the two NPA domains impaired water channel function, implicating associations of these domains with the formation of the aqueous pore. In this model, extracellular and cytoplasmic chamber domains with highly conserved NPA sequences are connected within the plasma membrane, forming a single aqueous pathway.

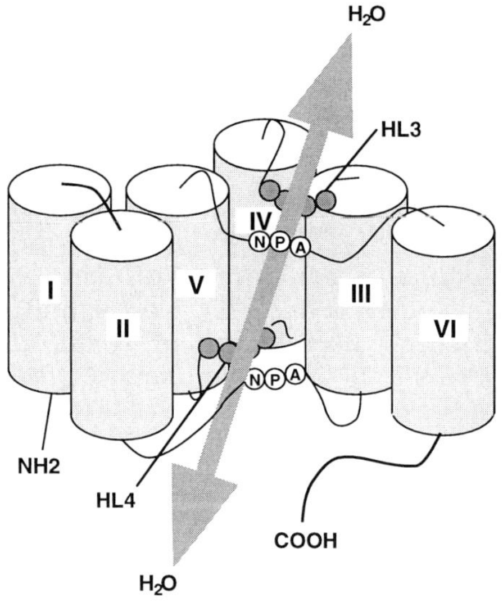

Figure 1. Three-dimensional structure model of AQP2. A central channel is formed between six transmembrane α-helices. Constrictions with a pore size of ~2 μ that determine selective water permeability is assembled with hydrophilic loop (HL)-3 and HL5 in the extracellular side, or HL2 and HL4 in the cytoplasmic side.

The structure of AQP2 was examined in our laboratory using similar strategy (14). In contrast to the results of AQP1, it was shown that the two chamber domains with NPA sequences are not critical for the formation of the aqueous pathway of AQP2 (Figure 1). From our findings, mutations near NPA boxes in the two NPA domains did not significantly alter water channel function and cystein residues inserted near the first NPA box did not induce alternative mercury-sensitive site. On the other hand, based on the findings that the replacement of the third hydrophilic loop (HL3) and the fourth hydrophilic loop (HL4) with hydrophobic residues decreased water permeability and that mercury sensitivity was found in these domains, it was suggested that the contribution of the HL3 and HL4 is significant in the formation of the aqueous pathway. On the basis of our results, we propose a structure model for AQP2 water channel. This structure model is compatible with recent cryo-electron microscopy studies for AQP1 (15, 16), which predicted a pore like structure assembled by six helices. The hydrophilic loops may be partially folded into the large pore, forming constrictions and hydrophilic apparatus that exclude charged ions and large molecules. These data require the revision of the original hourglass model, in which it is assumed that six transmembrane segments do not participate in the formation of the aqueous pore.

REGULATION OF AQUAPORIN FUNCTION

AQP2 and vasopressin antidiuresis

It has been believed that apical membrane water permeability of kidney collecting duct cells is regulated by vasopressin-dependent translocation of water-channel-containing vesicles via exo- and endocytosis between endosomal reservoir and the apical membrane. This is so-called the membrane shuttle hypothesis. After the identification of the vasopressin water channel AQP2, several lines of evidence provided by immunohistochemical studies have proved the hypothesis. The effects of vasopressin stimulation on the intracellular localization of AQP2 were demonstrated (17-19). Independently, the membrane shuttle hypothesis in toad bladder and rat renal papilla was confirmed by immunohistochemical and protein chemistry experiments (20, 21). Using V2 antagonist, it was shown that the distribution and quantity of AQP-CD protein in collecting duct principal cells are regulated by V2 signaling pathway (22). In addition, it was shown that long-term dehydration modified water permeability of the collecting duct cells. It has been known that water depletion for a long period increases the water permeability of the collecting duct even in the absence of vasopressin (23), and this phenomenon was explained by enhanced expression of AQP2 protein (22). Probably both persistent stimulation by vasopressin and increase in osmolarity of peritubuler interstitial space may stimulate the transcription of AQP2 messenger RNA.

To investigate the relation between V2 signaling and AQP2 function, effects of AQP2 phosphorylation by cAMP-dependent protein kinase have been examined. AQP2 has a consensus phosphorylation motif for cAMP-dependent protein kinase on the serine 256 near the carboxyl terminus. When expressed in Xenopus oocytes, phosphorylation of this site enhanced osmotic water permeability of AQP2 (24). However, direct regulation of AQP2 water permeability by its phosphorylation is still controversial (25). Effects of AQP2 phosphorylation on AQP2 trafficking were examined by expressing AQP2 in LLC-PK1 cells (26, 27). When a mutant lacking the serine 256 was expressed, vasopressin-dependent exocytosis of AQP2 was diminished assessed by functional assay and surface biotin labeling (Figure 2) (26). In addition, phosphorylation of S256A AQP2 was minimal in contrast to the enhanced phosphorylation of wild-type AQP2 after vasopressin stimulation (26). These data indicated that the serine 256 is required for cAMP-dependent regulatory exocytosis of AQP2. From accumulated observations, it is conceived that AQP2 recycling system is composed of cytoskeletal components (28-31), vesicle fusion proteins (32, 33), and AQP2 water channel. Vasopressin-responsive modification of actin (30, 34) and phosphorylation of several of the vesicle-associated proteins (35) have been shown. At present, it is believed that the translocation of AQP2 is regulated, at least in part, by phosphorylation of the channel itself.

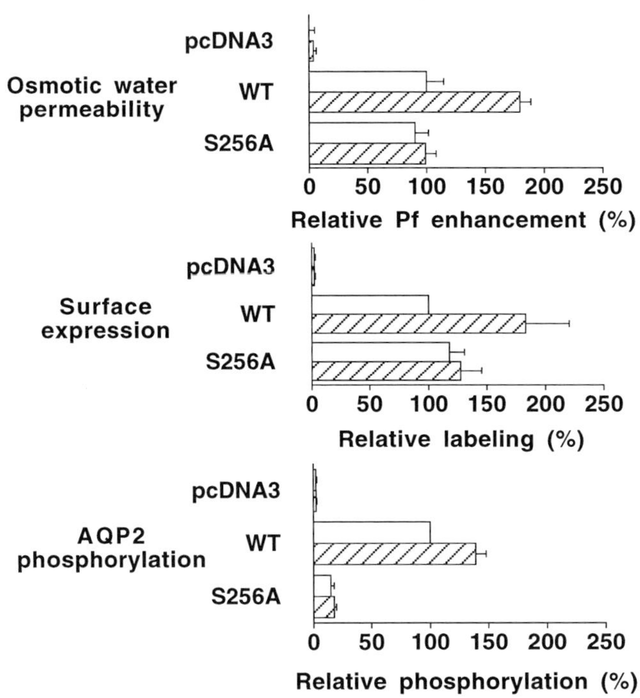

Figure 2. Comparison of osmotic water permeability (Pf), quantitative surface expression, and phosphorylation of wild-type (WT) and mutant AQP2. WT and a mutant with alanine in place of the serine 256 (S256A) were expressed in LLC-PK1 cells, and examined with (hatched bar) or without (open bar) cAMP stimulation. Data are shown in mean and SEM relative to WT without cAMP. Significant increase in Pf and surface expression by cAMP stimulation were noted in WT but not in S256A, showing that phosphorylation of the serine 256 is required for cAMP-dependent regulatory exocytosis of AQP2.

Regulations of other aquaporins

In contrast to AQP2, physiological regulation on channel translocation or water permeability for other AQP proteins has not been established. Regulatory ion conductance of AQP1 was reported (36), but not confirmed by other investigators (37-39). Although, MIP has been shown to be regulated by its phosphorylation (40) and to confer water permeability, effects of phosphorylation on water permeability are not known. γ-TIP is the first-identified aquaporin of plants (41), but regulatory mechanisms on γ-TIP is not clear. Enhancement of osmotic water permeability of γ -TIP by cAMP has been shown recently in Xenopus oocytes (10), but it is not clear whether translocation or gating of the channel is significant. Nodulin-26 was demonstrated as an ion channel (42) and cAMP-dependent phosphorylation was observed. Recently, nodulin-26 was shown to be an aquaporin (43), but effects of phosphorylation remain to be examined.

GENETIC DISORDERS OF AQUAPORINS

Nephrogenic diabetes insipidus and AQP2 mutation

A novel type of nephrogenic diabetes insipidus (NDI) was identified that have no mutation in V2 receptor gene but missense or nonsense mutations in AQP-2 gene (44, 45). The expression and characterization of equivalent mutants in Xenopus oocytes confirmed that these mutants are non-functioning. It was concluded that the patients do not respond to vasopressin due to these mutations, providing evidence that AQP2 is the vasopressin-regulated water channel. It is likely that many more mutants remain to be discovered. It was reported that when mutated AQP2 were expressed in Xenopus oocytes, channel protein was not transferred to the plasma membrane. To examine whether mistrafficking is the general cause of malfunction of AQP mutants, studies in mammalian epithelial cells are clearly needed. The possible molecular mechanisms for NDI include impairments in the formation of the water pore, in the maturation and the correct cellular localization of the channel protein, or in vasopressin-responsive channel trafficking.

AQP1 and null Colton blood type

Following the elucidation that Colton blood group antigen is associated with AQP1 (46), three individuals depleted of functional AQP1 water channel were found by searching null Colton blood type (47). They were shown to be homozygous for nonsense or missense mutations in AQP1 gene. Surprisingly, the three individuals were phenotypically normal with none or little functioning AQP1 in their RBC membranes and urine sediments. Although renal and other physiological status of the subjects were not reported in detail, the finding urged us to investigate the functional roles of the AQP1 water channel expressed in various tissues. Little is known about water channel physiology in tissues other than RBC and kidney, except recent observations that may suggest the physiological significance of the AQP1 water channel in lung alveolar cells (48), and cholangiocytes (49). It should be determined if the cell does not need the constitutive water pathway provided by AQP1, and if so, whether it is because diffusional water movement is enough for maintaining cell functions or because other transporters may compensate for the deficient water channel.

CONCLUSION

Aquaporins have been identified within last five years. Regulatory mechanisms of aquaporins have just become elucidated. Physiological relevance of many aquaporins remains to be clarified. In cell volume regulation, it is conceived that water movement is driven by solute gradient generated by several regulated ion channels. It is not known whether bulk water shift in regulatory volume change is mediated by aquaporins. Water movement through various channels or lipid bilayer may also be significant.

ACKNOWLEDGEMENT

This work was supported in part by Grant-in-Aid for Scientific Research (A) and (C) from the Ministry of Education, Science, Sports and Culture, Grant-in-Aid from The Japan Medical Association, and Grant-in-Aid from Tokyo Hypertension Conference.

REFERENCES

1 Fushimi K, Marumo F. Water channels. Curr Opin Nephrol Hypertens 1995; 4: 392-395.

2 Preston GM, Carroll TP, Guggino WB, Agre P. Appearance of water channels in Xenopus oocytes expressing red cell CHIP28 protein. Science 1992; 256: 385-387.

3 Fushimi K, Uchida S, Hara Y, Hirata Y, et al. Cloning and expression of apical membrane water channel of rat kidney collecting tubule. Nature 1993; 361: 549-552.

4 Ishibashi K, Sasaki S, Fushimi K, Uchida S, et al. Molecular cloning and expression of a member of the aquaporin family with permeability to glycerol and urea in addition to water expressed at the basolateral membrane of kidney collecting duct cells. Proc Natl Acad Sci U S A 1994; 91: 6269-6273.

5 Hasegawa H, Ma T, Skach W, Matthay MA, et al. Molecular cloning of a mercurial-insensitive water channel expressed in selected water-transporting tissues. J. Biol. Chem. 1994; 269: 5497-5500.
6 Raina S, Preston GM, Guggino WB, Agre P. Molecular cloning and characterization of an aquaporin cDNA from salivary, lacrimal, and respiratory tissues. J Biol Chem 1995; 270: 1908-1912.
7 Ishibashi K, Kuwahara M, Gu Y, Kageyama Y, et al. Cloning and functional expression of a new water channel abundantly expressed in the testis permeable to water, glycerol, and urea. J Biol Chem 1997; 272: 20782-20786.
8 Ishibashi K, Kuwahara M, Kageyama Y, Tohsaka A, et al. Cloning and functional expression of a second new aquaporin abundantly expressed in testis. Biochem Biophys Res Commun 1997; 237: 714-718.
9 Mulders SM, Preston GM, Deen PM, Guggino WB, et al. Water channel properties of major intrinsic protein of lens. J Biol Chem. 1995; 270: 9010-9016.
10 Maurel C, Chrispeels M, Lurin C, Tacnet F, et al. Function and regulation of seed aquaporins. J Exp Bot 1997; 48: 421-430.
11 Chrispeels MJ, Daniels MJ, Weig A. Aquaporins and water transport across the tonoplast. In: Leigh RA, Sanders D eds. Advances in Botanical Research. San Diego: Academic Press Inc, 1997; 25: 419-432.
12 Agre P, Preston GM, Smith BL, Jung JS, et al. Aquaporin CHIP: the archetypal molecular water channel. Am J Physiol 1993; F463-F476.
13 Jung JS, Preston GM, Smith BL, Guggino WB, et al. Molecular structure of the water channel through aquaporin CHIP. The hourglass model. J. Biol. Chem. 1994; 269: 14648-14654.
14 Fushimi K, Bai L, Marumo F, Sasaki S. Isolation of a gene endocing nodulin-like intrinsic protein of Esherichia coli. Biochem Mol. Biol. Intern. 1997; 41: 995-1003.
15 Walz T, Hirai T, Murata K, Heymann JB, et al. The three-dimensional structure of aquaporin-1. Nature 1997; 387: 624-627.
16 Cheng AC, vanHoek AN, Yeager M, Verkman AS, et al. Three-dimensional organization of a human water channel. Nature 1997; 387: 627-630.
17 Yamamoto T, Sasaki S, Fushimi K, Ishibashi K, et al. Vasopressin increases aqp-cd water channel in apical membrane of collecting duct cells in brattleboro rats. Am J Physiol 1995; 37: C1546-C1551.
18 Nielsen S, Chou CL, Marples D, Christensen EI, et al. Vasopressin increases water permeability of kidney collecting duct by inducing translocation of aquaporin-CD water channels to plasma membrane. Proc Natl Acad Sci U S A 1995; 92: 1013-1017.
19 DiGiovanni SR, Nielsen S, Christensen EI, Knepper MA. Regulation of collecting duct water channel expression by vasopressin in Brattleboro rat. Proc Natl Acad Sci U S A 1994; 91: 8984-8988.

20 Harris HJ, Zeidel ML, Jo I, Hammond TG. Characterization of purified endosomes containing the antidiuretic hormone-sensitive water channel from rat renal papilla. J Biol Chem 1994; 269: 11993-12000.
21 Paredes A, Briscoe DM, Williams CK, Garcia PA, et al. Water channel vesicles from toad urinary bladder contain a family of proteins present in other tissues. Am J Physiol 1994; C1366-C1375.
22 Hayashi M, Sasaki S, Tsuganezawa H, Monkawa T, et al. Expression and distribution of aquaporin of collecting duct are regulated by vasopressin V2 receptor in rat kidney. J Clin Invest 1994; 94: 1778-1783.
23 Wade JB, Nielsen S, Coleman RA, Knepper MA. Long-term regulation of collecting duct water permeability: freeze- fracture analysis of isolated perfused tubules. Am J Physiol 1994; F723-F730.
24 Kuwahara M, Fushimi K, Terada Y, Marumo F, et al. cAMP-dependent phosphorylation stimulates water permeability of aquaporin-collecting duct water channel protein expressed in Xenopus oocytes. J Biol Chem 1995; 270: 10384-10387.
25 Lande MB, Jo I, Zeidel ML, Somers M, et al. Phosphorylation of aquaporin-2 does not alter the membrane water permeability of rat papillary water channel-containing vesicles. J Biol Chem 1996; 271: 5552-5557.
26 Fushimi K, Sasaki S, Marumo F. Phosphorylation of serine 256 is required for cAMP-dependent regulatory exocytosis of the aquaporin-2 water channel. J Biol Chem 1997; 272: 14800-14804.
27 Katsura T, Gustafson CE, Ausiello DA, Brown D. Protein kinase A phosphorylation is involved in regulated exocytosis of aquaporin-2 in transfected LLC-PK1 cells. Am J Physiol 1997; 41: F816-F822.
28 Sabolic I, Katsura T, Verbavatz JM, Brown D. The AQP2 water channel: effect of vasopressin treatment, microtubule disruption, and distribution in neonatal rats. J Mem Biol 1995; 143: 165-175.
29 Holmgren K, Magnusson KE, Franki N, Hays RM. ADH-induced depolymerization of F-actin in the toad bladder granular cell: a confocal microscope study. Am J of Physiol 1992; 262: C672-C677.
30 Simon H, Gao Y, Franki N, Hays RM. Vasopressin depolymerizes apical F-actin in rat inner medullary collecting duct. Am J Physiol 1993; 265: C757-C762.
31 Verkman AS, Weyer P, Brown D, Ausiello DA. Functional water channels are present in clathrin-coated vesicles from bovine kidney but not from brain. J Biol Chem 1989; 264: 20608-200613.
32 Franki N, Macaluso F, Gao Y, Hays RM. Vesicle fusion proteins in rat inner medullary collecting duct and amphibian bladder. Am J Physiol 1995; 268: C792-C797.
33 Franki N, Macaluso F, Schubert W, Gunther L, et al. Water channel-carrying vesicles in the rat imcd contain cellubrevin. Am J Physiol 1995; 38: C797-C801.

34 Ding GH, Franki N, Condeelis J, Hays RM. Vasopressin depolymerizes F-actin in toad bladder epithelial cells. Am J Physiol 1991; 260: C9-C16.

35 Harris HW Jr. Antidiuretic hormone modulates membrane phosphoproteins in toad urinary bladder and retrieved water channel containing apical membrane vesicles. J Am Soc Nephrol 1991; 1: 1114-1122.

36 Yool AJ, Stamer WD, Regan JW. Forskolin stimulation of water and cation permeability in aquaporin 1 water channels. Science 1996; 273: 1216-1218.

37 Sasaki S, Uchida S, Kuwahara M, Fushimi K, et al. Aquaporins and ion conductance. Science 1997; 275: 1490-1491.

38 Agre P, Lee MD, Devidas S, Guggino WB. Aquaporins and ion conductance. Science 1997; 275: 1490-1490.

39 Verkman AS, Yang BX. Aquaporins and ion conductance. Science 1997; 275: 1491-1491.

40 Ehring GR, Lagos N, Zampighi GA, Hall JE. Phosphorylation modulates the voltage dependence of channels reconstituted from the major intrinsic protein of lens fiber membranes. J Memb Biol 1992; 126: 75-88.

41 Maurel C, Reizer J, Schroeder JI, Chrispeels MJ. The vacuolar membrane protein γ-TIP creates water specific channels in Xenopus oocytes. Embo J 1993; 12: 2241-2247.

42 Weaver CD, Shomer NH, Louis CF, Roberts DM. Nodulin 26, a nodule-specific symbiosome membrane protein from soybean, is an ion channel. J. Biol. Chem. 1994; 269: 17858-17862.

43 Rivers RL, Dean RM, Chandy G, Hall JE, et al. Functional analysis of nodulin 26, an aquaporin in soybean root nodule symbiosomes. J Biol Chem 1997; 272: 16256-16261.

44 Deen PM, Verdijk MA, Knoers NV, Wieringa B, et al. Requirement of human renal water channel aquaporin-2 for vasopressin- dependent concentration of urine. Science 1994; 264: 92-95.

45 van Lieburg AF, Verdijk MA, Knoers VV, van Essen AJ, et al. Patients with autosomal nephrogenic diabetes insipidus homozygous for mutations in the aquaporin 2 water-channel gene. Am J Hum Genet 1994; 55: 648-652.

46 Smith BL, Preston GM, Spring FA, Anstee DJ, et al. Human red cell aquaporin CHIP. I. Molecular characterization of ABH and Colton blood group antigens. J Clin Invest 1994; 94: 1043-1049.

47 Preston GM, Smith BL, Zeidel ML, Moulds JJ, et al. Mutations in aquaporin-1 in phenotypically normal humans without functional CHIP water channels. Science 1994; 265: 1585-1587.

48 Folkesson HG, Matthay MA, Hasegawa H, Kheradmand F, et al. Transcellular water transport in lung alveolar epithelium through mercury-sensitive water channels. Proc Natl Acad Sci U S A 1994; 91: 4970-4974.

49 Roberts SK, Yano M, Ueno Y, Pham L, et al. Cholangiocytes express the aquaporin CHIP and transport water via a channel-mediated mechanism. Proc Natl Acad Sci U S A 1994; 91: 13009-13013.

Cell Volume Regulation: The Molecular Mechanism and Volume Sensing Machinery
Y. Okada, editor.

Water movement during cell volume regulation

Shigeru Morishima[a], Hajime Kida, Shunji Ueda[c], Tsutomu Chiba[d], Yasunobu Okada

[a]Department of Cellular and Molecular Physiology, National Institute for Physiological Sciences, Okazaki 444-8585

[b]Department of Gastroenterological Endoscopy, Faculty of Medicine, Kyoto University, Kyoto 606-8507, Japan

[c]Department of Internal Medicine, Faculty of Medicine, Kyoto University, Kyoto 606-8507, Japan

[d]Department of Gastroenterology, Faculty of Medicine, Kyoto University, Kyoto 606-8507, Japan

INTRODUCTION

Most kinds of cells have a volume regulatory function, and a variety of ion channels and transporters are known to be activated and involved in volume regulation [1]. Although the osmotic balance is maintained by the movement of ionic and nonionic osmolytes across the cell membrane, little is known about water movement when cells are exposed to different osmotic conditions. We have recently started the study on the mechanism of water transport of a cultured cell line, Intestine 407, which originated from human intestine epithelium, during hypo- and hypertonic stress using the fast image analysis system.

MEASUREMENT OF CELL VOLUME

Water accounts for the most part of the weight of a cell, and it is also believed that the volume of a cell is roughly proportional to the amount of water the cell has in it. For that reason, to measure the water movement across the membrane, it is essential first to measure the cell volume. Cell volume measurements have been done by means of imaging analysis, light deflection method, Coulter™ counter, and so on [2]. Among them, the image analysis method has been frequently adopted to detect the volume change of a single cell as the others can only provide the mean value of data from a large number of cells. Recently the single-cell image analysis technique was frequently used for the screening of functional expression of cloned water channel gene in *Xenopus* oocytes, the diameter of which reaches 1 mm. As they are spherical and proof against the microscopic light, we can get the round shadow in the microscopic image. It is easy to analyze the image and to calculate the volume from the diameter. However, it is much more difficult to measure the single cell volume of an "ordinal" mammalian cell. Sometimes cells have irregular shapes, and even if the shape looks spherical, like a culture cell line we have been using, it is still difficult to determine the edge and, as a result, also the diameter of a cell. This edge determination has been mostly done by eye. The weakest point of this method is its slow processing speed. If one really wants to analyze the water movement and calculate the osmotic permeability coefficient (*Pf*), one should closely examine the volume change rapidly on the order of seconds. If we sample an image every 1 sec for 30 min, then the total count of images that we must calculate the cell volume is 1800. It *is* a painful effort to repeat the image analysis by eye.

Recently we are developing the automatic edge determination system composing of a CCD camera, an image capture adapter and a personal computer. This high-speed automatic cell image analysis system makes it possible to record the image every 1/30 sec and to determine the volume change automatically even if the recorded images were more than a thousand.

Figure 1 shows examples of actual measurements in Intestine 407 cells exposed to a hypotonic or hypertonic solution. Both volume changes shortly after an osmotic challenge and following slow processes of volume regulation were consistently observed.

Figure 1. Volume regulation of Intestine407 cells measured by high-speed automatic image analysis. Cells are exposed to 80% hypotonic solution (A) and 30% hypertonic solution at 37 °C (B).

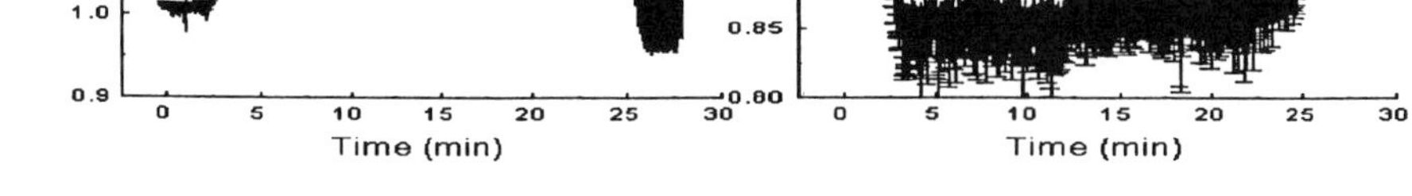

WATER INFLOW SHORTLY AFTER THE EXPOSURE TO HYPOTONIC SOLUTION

When cells are exposed to a hypotonic solution, cells suddenly start to expand. Our first questions were: How much does water enter a cell? When does entrance of water stop? It has been believed that a cell continues expanding and absorbing the water until its cytoplasmic osmolality equilibrates with that of extracellular solution. When much water comes into a cell, hydrostatic pressure increases; and if the cell membrane cannot support its pressure, it will be ruptured. Which is more important to stop the water inflow, the hydrostatic pressure or the cytoplasmic osmolality? To answer these questions, we measured the volume expansion perfusing the external solution of different osmolality. The maximum volume that cells reach after a hypotonic challenge was plotted against the osmolality (Fig. 2) and proved to be on a straight line. This implies that the osmotic pressure inside the cells (Π) satisfies the following equation:

$$\Pi(V-b)=\Pi_0(V_0-b)$$

where V is the maximum cell volume after hypotonic challenge, V_0 is the initial cell volume in isotonic solution of osmolality Π_0, and b is the non-water fraction of cell volume [3]. From this result, we concluded that the cytoplasmic osmolality, rather than increased hydrostatic pressure, regulates the water inflow.

Figure 2. **Maximum cell volume in different osmolar solutions.** Cells were perfused to hypotonic solutions of different osmolality, and the ration of maximum volume to control is plotted against the inverse of the osmolality. Data represent the mean ± SEM of 5 to 6 experiments.

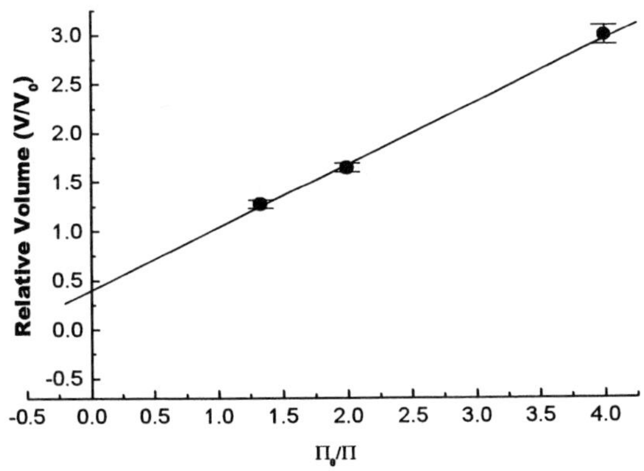

TRIGGER THAT STARTS THE VOLUME REGULATION

After osmotic swelling, most cell types including Intestine 407 cells exhibit a regulatory volume decrease (RVD) [1]. One may imagine that the disruption of cytoskeleton may start the volume regulation, but little is known about the real trigger that induces the volume regulation. Is the increase of the pressure of cytoplasmic side or the tension of membrane important? RVD occurs after some delay of cell expansion. It is difficult simply to determine this delay and the time when exactly RVD starts. However, we did not observe a significant difference in the delay among hypotonic challenges with different osmolality. So, we can safely assume that neither hydrostatic pressure nor tension of membrane defines the starting point of RVD.

EXISTENCE OF WATER CHANNEL AND ITS ROLE ON VOLUME EXPANSION AND REGULATION?

Upon the osmotic stress and during the following volume regulatory period, water must move into or out of cells quickly. Which pathway can water pass through? During the volume regulation, cells should be sensitive to the subtle difference of osmolality, as only a small amount of ions or osmolytes should be rapidly followed by the water movement. Does it pass through aquaporin water channels [4] or go across the lipid bilayer? There has been also much discussion whether a variety of ionic channels, especially water-filled single-filed channels, can permeate water [5]. To answer this question, the measurements of the *Pf* value and its sensitivity to temperature as well as the ratio of *Pf* to the diffusion permeability coefficient *Pd* are prerequisite. Using a high-speed automatic cell image analysis system, an investigation along these lines is currently being undertaken in our laboratory.

REFERENCES

1 Okada Y. Volume expansion-sensing outward-rectifier Cl^- channel: fresh start to the molecule identity and volume sensor. *Am J Physiol* 1997; 273. C755-C789.

2. Verkman AS, van Hoek AN, Ma, T., Frigeri A, Skach WR, Mitra A, Tamarappoo BK, Farinas J. Water transport across mammalian cell membranes. *Am J Physiol* 1996; 270. C12-30.

3. Sachs JR. Cell volume regulation. In: Schutz SG, Andreoli, TE, Brown A, Fambrough DM, Hoffman JF, Welsh, MJ, eds. *Molecular biology of membrane transport disorders.* New York: Plenum, 1996;

4. van Os CH, Deen PM, Dempster JA. Aquaporins: water selective channels in biological membranes. Molecular structure and tissue distribution. *Biochim et Biophys Acta (Rev Biomembr)* 1994; 1197. 291-309.

5. Sabirov RZ, Morishima S, Okada Y. Probing the water permeability of ROMK1 and amphotericin B channels using *Xenopus* oocytes. *Biochim Biophys Acta* 1997; 1368. 19-26

Index of authors

Anzai, N., 151

Bazzini, C., 131

Cardin, V., 43
Chen, L., 109
Chiba, T., 209

De Jonge, H.R., 59
Dienstl, A., 131
Droogmans, G., 141

Eggermont, J., 141
Enomoto, K., 85

Fujita, T., 159
Fujiwara, K., 23
Fürst, J., 131
Furuya, K., 85
Fushimi, K., 199

Gschwentner, M., 131
Gulbins, E., 49

Harada, N., 23
Hazama, A., 15, 93
Hirose, S., 159
Hohlrieder, M., 131

Ikeda, T., 167
Ito, Y., 67
Iwamoto, T., 167
Izumida, I., 151

Jacob, T.J.C., 109

Kawahara, K., 151
Kida, H., 209
Kimelberg, H.K., 189
Kimura, C., 67
Kobayashi, Y., 167
Komuro, I., 99
Koyama, T., 67
Kubo, M., 125
Kusano, K., 23
Kwon, H.M., 175

Lang, F., 49
Lepple-Wienhues, A., 49
Lopes, R.B., 23

Marumo, F., 199
Masuda, M., 23
Mitsuiye, T., 35
Miwa, A., 15, 93
Miyoshi, T., 15, 93
Morales-Mulia, S., 43
Morishima, S., 15, 209

Nakamura, N., 159
Nakanishi, T., 181
Nakano, H., 85
Nilius, B., 141
Noma, A., 35

Oehl, B., 131
Oike, M., 67
Oiki, S., 15, 125
Okada, Y., 15, 93, 125, 209
Osawa, M., 23

Pasantes-Morales, H., 43
Paulmichl, M., 131

Quesada, O., 43

Ritter, M., 49, 131
Rutledge, E.M., 189

Sachs, J.R., 3
Sakuta, H., 159
Sasaki, S., 199
Scandella, E., 131
Schmarda, A., 131
Shigekawa, M., 167
Shigematsu, H., 23
Shimizu, T., 93
Sokabe, M., 71
Suzuki, Y., 159
Szabo, I., 49

Takamitsu, Y., 181
Takano, K., 159
Takemoto, K., 71
Tilly, B.C., 59
Tominaga, M., 15
Tominaga, T., 15
Tsumura, T., 15

Ueda, K., 15
Ueda, S., 209

Van der Wijk, T., 59
Voets, T., 141

Wakabayashi, S., 167
Waldegger, S., 49
Wang, L., 109
Wöll, E., 131

Yamamoto, S., 181
Yamauchi, A., 181
Yu, W., 71